Selected Titles in This Series

(Continued in the back of this publication)

Sub-Laplacians with Drift on Lie Groups of Polynomial Volume Growth

Number 739

Sub-Laplacians with Drift on Lie Groups of Polynomial Volume Growth

Georgios K. Alexopoulos

January 2002 • Volume 155 • Number 739 (end of volume) • ISSN 0065-9266

American Mathematical Society
Providence, Rhode Island

2000 *Mathematics Subject Classification.*
Primary 22E15, 22E25, 22E30, 43A80.

Library of Congress Cataloging-in-Publication Data

Alexopoulos, Georgios K., 1962–
Sub-Laplacians with drift on Lie groups of polynomial volume growth / Georgios K. Alexopoulos.
p. cm. — (Memoirs of the American Mathematical Society, ISSN 0065-9266 ; no. 739)
"Volume 155, number 739 (end of volume)."
Includes bibliographical references.
ISBN 0-8218-2764-2 (alk. paper)
1. Lie groups. 2. Functional analysis. I. Title. II. Series.

QA3 .A57 no. 739
[QA387]
510 s—dc21
[512′.55] 2001045834

Memoirs of the American Mathematical Society

This journal is devoted entirely to research in pure and applied mathematics.

Subscription information. The 2002 subscription begins with volume 155 and consists of six mailings, each containing one or more numbers. Subscription prices for 2002 are \$524 list, \$419 institutional member. A late charge of 10% of the subscription price will be imposed on orders received from nonmembers after January 1 of the subscription year. Subscribers outside the United States and India must pay a postage surcharge of \$31; subscribers in India must pay a postage surcharge of \$43. Expedited delivery to destinations in North America \$35; elsewhere \$130. Each number may be ordered separately; *please specify number* when ordering an individual number. For prices and titles of recently released numbers, see the New Publications sections of the *Notices of the American Mathematical Society*.

Back number information. For back issues see the *AMS Catalog of Publications*.

Subscriptions and orders should be addressed to the American Mathematical Society, P. O. Box 845904, Boston, MA 02284-5904. *All orders must be accompanied by payment.* Other correspondence should be addressed to Box 6248, Providence, RI 02940-6248.

Memoirs of the American Mathematical Society is published bimonthly (each volume consisting usually of more than one number) by the American Mathematical Society at 201 Charles Street, Providence, RI 02904-2294. Periodicals postage paid at Providence, RI. Postmaster: Send address changes to Memoirs, American Mathematical Society, P. O. Box 6248, Providence, RI 02940-6248.

This publication is indexed in *Science Citation Index*®, *SciSearch*®, *Research Alert*®, *CompuMath Citation Index*®, *Current Contents*®*/Physical, Chemical & Earth Sciences.*
Printed in the United States of America.

♾ The paper used in this book is acid-free and falls within the guidelines established to ensure permanence and durability.
Visit the AMS home page at URL: `http://www.ams.org/`

10 9 8 7 6 5 4 3 2 1 07 06 05 04 03 02

Contents

Abstract

We prove a parabolic Harnack inequality for a centered sub-Laplacian L on a connected Lie group G of polynomial volume growth by using ideas from Homogenisation theory and by adapting the method of Krylov and Safonov. We use this inequality to obtain a Taylor formula for the heat functions and thus we also obtain Harnack inequalities for their space and time derivatives. We characterise the harmonic functions which grow polynomially. We obtain Gaussian estimates for the heat kernel and estimates similar to the classical Berry-Esseen estimate. Finally, we study the associated Riesz transform operators. If L is not centered, then we can conjugate L by a convenient multiplicative function and obtain another centered sub-Laplacian L_C. Thus our results also extend to non-centered sub-Laplacians .

2000 *Mathematics Subject Classification.* 22E15, 22E25, 22E30, 43A80.

Key words and phrases. Lie group, volume growth, sub-Laplacian, drift, homogenised operator, harmonic function, Harnack inequality, heat kernel, Riesz transform.

1. Introduction and statement of the results

Let G be a connected Lie group, let dg be a left invariant Haar measure on G and let $|A| = dg$-measure(A), A being a Borel measurable subset of G.

Let us fix a compact neighborhood U of the identity element e of G and let

$$U^n = \{g_1 g_2 ... g_n : g_1, g_2, ..., g_n \in U\}.$$

We set

$$|g|_G = \inf\{n : g \in U^n\}. \tag{1.1}$$

We shall assume that G has polynomial volume growth, i.e. there is a constant $c > 0$ such that $|U^n| \leq cn^c, n \in \mathbb{N}$. Then, (cf. [**Gui**]) there is an integer $D \geq 0$, such that

$$\frac{1}{c} n^D \leq |U^n| \leq cn^D, \quad n \in \mathbb{N}. \tag{1.2}$$

We call D homogeneous dimension (at infinity) of G. Note that D does not depend on the choice of U. Note also that every connected nilpotent Lie group has polynomial volume growth.

We identify the Lie algebra $\mathfrak{g}$ of G with the left invariant vector fields on G.

By a left invariant sub-Laplacian on G, we shall understand an operator

$$L = -(E_1^2 + ... + E_p^2) + E_0$$

where $E_0, E_1, ..., E_p$ are left invariant vector fields and where the vector fields $E_1, ..., E_p$ satisfy Hörmander's condition, i.e. they generate together with their successive Lie brackets $[E_{i_1}, [E_{i_2}, [..., E_{i_k}]...]]], 1 \leq i_j \leq p, 1 \leq j \leq k$ the Lie algebra $\mathfrak{g}$ of G.

For results concerning the analysis of the symmetric sub-Laplacians (i.e. with drift term $E_0 = 0$) we reffer the reader to [**VSC**] (and the refferences therein).

In this paper, we are mainly interested in non-symmetric sub-Laplacians (i.e. with $E_0 \neq 0$). We are interested, in particular, in the large time behavior of the heat kernel $p_t(x, y)$ of L (i.e. the fundamental solution of the equation $(\frac{\partial}{\partial t} + L)u = 0$). Most of the techniques developped in [**VSC**] work only for symmetric operators. So the presence of a drift term adds a new essential difficulty to the problem.

Our interest is also motivated by the central limit theorem. If we restrict our interest to symmetric sub-Laplacians, then we can only study products of symmetric random variables (with values on G). In order to deal with non-symmetric random variables, we need to understand the large time behavior of the heat kernel of non-symmetric sub-Laplacians.

Throughout this article, by the term simply connected we shall understand connected and simply connected.

The different constants will always be denoted with the same letter c. When their dependence or idependence is significant, it will be clearly stated.

Finally, we shall denote by $\mathfrak{e}_1$ the linear sub-space of $\mathfrak{g}$ generated by the vector fields $E_1, ..., E_p$ and by $\mathfrak{e}_2$ the linear sub-space of $\mathfrak{g}$ generated by the Lie brackets $[E_i, E_j]$, $1 \leq i, j \leq p$.

Received by the editor February 10, 1997, and in revised form November 27, 1998.

ACKNOWLEDGEMENT. I wish to thank Nikolaos Varopoulos and Noël Lohoué for the encouragement to this research and for several inspiring discussions during the preparation of this article.

1.1. The general strategy. Our approach is inspired by Homogenisation theory (cf. [**BLP, JKO**]). We can prove for example that every left invariant sub-Laplacian L is behaving at infinity like another, so called, homogenised sub-Laplacian L_H which lives on a nilpotent Lie group and hence it is much easier to deal with.

The existence of the homogenised sub-Laplacian L_H allows us to distinguish those drifts whose large time effect is not important. Hence the notion of a centered sub-Laplacian.

The main tool is a parabolic Harnack inequality which holds only for large time (or for large balls in the elliptic case). It is proved by adaptating the method of Krylov and Safonov (cf. [**KS**]).

Note that as it was shown by Varopoulos [**V2, VSC**], the method of Moser [**M1, M2**] can also be adapted to obtain Harnack inequalities in abstract contexts. In our case though, Moser's approach seems dificult to adapt.

The reader can observe that the same proof also works for convolution powers of densities on connected Lie groups as well as for convolution powers of probability measures on discrete groups of polynomial volume growth.

Once we have a parabolic Harnack inequality, then we can prove a large time Gaussian estimate for the heat kernel $p_t(x,y)$ by some elementary method.

We adapt, in the context of Lie groups, some ideas of Bergström (cf. [**Be, Saz**]) who proves the central limit theorem by the so-called convolution method. This method does not make use of the characteristic functions. It gives also simultaneously the speed of convergence (which according to the classical Berry-Esseen theorem is $1/\sqrt{n}$). By this method, once we have a Harnack inequality, we automatically have a Berry-Esseen estimate for the $p_t(x,y)$ (although not at the optimal rate of convergence).

In order to controll the derivatives of the heat functions (i.e. the functions u satisfying $\left(\frac{\partial}{\partial t}+L\right)u=0$) we proceed as follows. We observe that the Berry-Esseen estimate provides with a certain compactness the space of the heat functions. Using this compactness, we can adapt some ideas of Avellaneda and Lin [**AL1, AL2**] (see also [**A1, A2, ALo**]) and prove a Taylor formula for the heat functions. As an immediate consequence of this formula, we have Harnack inequalities for the space and time derivatives of the positive heat functions.

We obtain the optimal rate of convergence for Berry-Esseen estimate for the $p_t(x,y)$. We also obtain Berry-Esseen estimates for the time and space derivatives of the $p_t(x,y)$.

Finally, we study the associated Riesz trasform operators by using the Berry-Esseen estimate for the spacial derivatives of the $p_t(x,y)$.

1.2. Centered sub-Laplacians. Let G_1 be the closure of $[G,G]$ in G. Then, G/G_1 is a connected abelian Lie group and hence it can de written as a direct product $\mathbb{R}^n\times\mathbb{T}^m$, where $\mathbb{T}=\mathbb{R}/\mathbb{Z}$. Let us consider the canonical projection $\pi: G\to\mathbb{R}^n$, let $H=\mathrm{Ker}(\pi)$ and let $\mathfrak{h}$ be the Lie algebra of H. Then $\mathfrak{h}$ is an ideal of $\mathfrak{g}$ and $[\mathfrak{g},\mathfrak{g}]\subseteq\mathfrak{h}$.

We shall say that the sub-Laplacian L is centered if $E_0\in\mathfrak{h}$.

1. Intuitive justification. When the drift term $E_0 \in \mathfrak{h}$ then its large time effect is not important. For example, we can obtain a Gaussian estimate for the heat kernel similar to the one known in the case when $E_0 = 0$ (cf. [**VSC**]).

More precisely, let $L_S = -(E_1^2 + ... + E_p^2)$ and let $p_t^S(x, y)$ be the heat kernel of L_S. Then there is $c > 0$ suct that (cf. [**V1, V3, SC, VSC**]) for all $x, y \in G$, $t \geq 1$

$$
\begin{aligned}
p_t^S(x,y) &\leq ct^{-D/2} \exp\left(-\frac{|x^{-1}y|_G^2}{ct}\right) \\
p_t^S(x,y) &\geq \frac{1}{c} t^{-D/2} \exp\left(-c\frac{|x^{-1}y|_G^2}{t}\right).
\end{aligned} \tag{1.2.1}
$$

We shall prove that the above estimates are still satisfied by the heat kernel $p_t(x, y)$ of $L = L_S + E_0$, when $E_0 \in \mathfrak{h}$.

Intuitively, this can be explained as follows. Let us pretend, for simplification, that E_0 is also in the center of $\mathfrak{g}$. Then, we have that $p_t(x, e) = p_t^S(\exp(tE_0)x, e)$. So the existence of the drift term E_0 implies that most of the mass of $p_t(x, e)$ lies in a ball of radius $\sqrt{t}$ and centered at the point $\exp(-tE_0)$ instead a ball of radius $\sqrt{t}$ and centered at the identity element e of G. If $E_0 \in \mathfrak{h}$, then, grossly speaking, we have to deal with some of following three cases:

1. E_0 is in the Lie algebra of a compact Lie subgroup of G. Then there is $c > 0$ such that
$$|\exp(tE_0)|_G \leq c, \qquad t \in \mathbb{R}$$
and hence the estimates (1.2.1) are still valid.
2. G is nilpotent and $E_0 \in [\mathfrak{g}, \mathfrak{g}]$ (take for example the case of the Heisenberg group). Then there is $c > 0$ such that
$$|\exp(tE_0)|_G \leq c\sqrt{|t|}, \qquad |t| \geq 1.$$
Hence, the estimates (1.2.1) are still valid.
3. This third case occurs when G is not nilpotent. The simplest example is the the group of Euclidean motions on the plane. This group is isomorphic to the semidirect product $\mathbb{R}^2 \rtimes \mathbb{T}$, where $\mathbb{T}$ acts on $\mathbb{R}^2$ by rotations. Note that then $H = \mathbb{R}^2 \rtimes \mathbb{T}$. The drift E_0 splits into two parts $E_0 = E_{0,\mathbb{R}^2} + E_{0,\mathbb{T}}$. The part $E_{0,\mathbb{T}}$ lies in $\mathbb{T}$ which is compact and hence by the case (1) has no effect on the estimates (1.2.1). To treat the other term $E_{0,\mathbb{R}^2}$ which lies in $\mathbb{R}^2$, we observe that L can be written as an operator on $\mathbb{R}^2 \times \mathbb{T}$. Then, $E_{0,\mathbb{R}^2}$ is a vector field with periodic coefficients in $\mathbb{R}^2$ (cf. [**A1, A2**]). So, what influences the large time behavior of the heat kernel, is not the drift term $E_{0,\mathbb{R}^2}$ itself, but the "average" or effective drift (cf. [**JKO** p. 68]). In this particular example the effective drift vanishes. In the general case the effective drift lies in the first commutator of some nilpotent Lie group and hence, by (2), has no effect in the Gaussian estimates (1.2.1).

1.3. The passage from a non-centered to a centered sub-Laplacian. If L is not centered, then by conjugating L with a convenient multiplicative function on G we can obtain a centered sub-Laplacian. Thus, a question about a non-centered sub-Laplacian can be rephrased to a question about a centered one. More presicely, let $L = -\left(E_1^2 + ... + E_p^2\right) + E_0$ be a left invariant sub-Laplacian on a connected Lie group G of polynomial volume growth and let us assume that L is not centered, i.e. $E_0 \notin \mathfrak{h}$.

By a multiplicative functon χ on G, we shall understand a function $\chi : G \to \mathbb{R}^+$ satisfying $\chi(xy) = \chi(x)\chi(y)$, $x, y \in G$. Note that then χ can be written as

$$\chi = \phi \circ \pi \quad \text{with} \quad \phi(x) = e^{\langle b,x\rangle}$$

where π is the quotient map $\pi : G \to \mathbb{R}^n \cong G/H$ and where $\langle b, x\rangle = b_1x_1 + ... + b_kx_k$ for $b = (b_1, ..., b_k), x = (x_1, ..., x_k) \in \mathbb{R}^k$.

We have the following well known lemma (see for example [**MV**]):

LEMMA 1.3.1. *Let L be as above. Then, there is a constant $\beta > 0$, a vector field $Y \in \mathfrak{h}$ and a multiplicative function χ such that*

$$L = \chi^{-1} (L_C + \beta) \chi, \tag{1.3.1}$$

where

$$L_C = -\left(E_1^2 + ... + E_p^2\right) + Y.$$

If $p_t^C(x, y)$ is the heat kernel of L_C, then it follows from the above lemma that

$$p_t(x, y) = e^{-\beta t}\chi(x^{-1})p_t^C(x, y)\chi(y), \quad x, y \in G, t > 0 \tag{1.3.2}$$

For reasons of completeness, we give below the proof of the above lemma.

PROOF. Let π the quotient map $\pi : G \to \mathbb{R}^k \cong G/H$ and let us chose a basis $\{X_1, ..., X_q\}$ of $\mathfrak{g}$ such that

$$\{X_{n+1}, ..., X_q\} \subseteq \mathfrak{h} \quad \text{and} \quad d\pi(X_i) = \frac{\partial}{\partial x_i}, \quad 1 \leq i \leq n.$$

Then L can also be written as

$$L = -\sum_{1\leq i,j\leq q} a_{ij}X_iX_j + \sum_{1\leq i\leq q} a_iX_i$$

with $a_{ij} = a_{ji}, 1 \leq i, j \leq q$.

It follows that the image $d\pi(L)$ of L can be written as

$$d\pi(L) = -\sum_{1\leq i,j\leq n} a_{ij}\frac{\partial}{\partial x_i}\frac{\partial}{\partial x_j} + \sum_{1\leq i\leq n} a_i\frac{\partial}{\partial x_i}.$$

Note that the $(n \times n)$ matrix $B = (b_{ij})$ with entries $b_{ij} = a_{ij}$, $1 \leq i, j \leq n$ is positive definite.

Let $f \in C^\infty$, let $\chi(x) = e^{\langle b, \pi(x)\rangle}$ be a multiplicative function on G and let $a = (a_1, ..., a_n)$ and $b = (b_1, ..., b_n)$. Then

$$\begin{aligned} L(\chi f) = & - \sum_{1\le i,j\le q} a_{ij}\chi X_i X_j f - \sum_{1\le i,j\le q} a_{ij}(X_i\chi)X_j f - \sum_{1\le i,j\le q} a_{ij}(X_i f)X_j\chi \\ & + \sum_{1\le i\le q} a_i\chi X_i(f) - \sum_{1\le i,j\le q} a_{ij} f X_i X_j \chi + \sum_{1\le i\le q} a_i f X_i\chi \\ = & - \chi \sum_{1\le i,j\le q} a_{ij} X_i X_j f - \chi \sum_{\substack{1\le i\le n \\ 1\le j\le q}} a_{ij} b_i X_j(f) - \chi \sum_{\substack{1\le i\le q \\ 1\le j\le n}} a_{ij} b_j X_i f \\ & + \chi \sum_{1\le i\le q} a_i X_i f - f\chi \sum_{1\le i,j\le n} a_{ij} b_i b_j + f\chi \sum_{1\le i\le n} a_i b_i \\ = & - \chi \sum_{1\le i,j\le q} a_{ij} X_i X_j f + \chi \sum_{n<i\le q} \left(a_i - 2 \sum_{1\le j\le n} a_{ij} b_j \right) X_i f \\ & + \chi \sum_{1\le i\le n} \left(a_i - 2\,(Bb)_i\right) X_i f - \left(\langle b, Bb\rangle - \langle b, a\rangle\right) f\chi. \end{aligned}$$

Hence

$$\begin{aligned} \chi^{-1} L\chi = - \left(E_1^2 + ... + E_p^2\right) + \sum_{n<i\le q} \left(a_i - 2 \sum_{1\le j\le n} a_{ij} b_j \right) X_i \\ + \sum_{1\le i\le n} \left(a_i - 2\,(Bb)_i\right) X_i - \left(\langle b, Bb\rangle - \langle b, a\rangle\right) \end{aligned}$$

We want to have

$$a_i - 2\,(Bb)_i = 0, \qquad 1 \le i \le n$$

or else that

$$a = 2Bb.$$

Since the matrix B is invertible, this is equivalent to

$$b = \frac{1}{2} B^{-1} a$$

For this choice of b the constant term becomes

$$\beta = - \left(\langle b, Bb\rangle - \langle b, a\rangle\right) = \langle b, Bb\rangle > 0.$$

Also, if we set

$$Y = \sum_{n<i\le q} \left(a_i - 2 \sum_{1\le j\le n} a_{ij} b_j \right) X_i,$$

then $Y \in \mathfrak{h}$ and

$$\chi^{-1} L\chi = - \left(E_1^2 + ... + E_p^2\right) + Y + \beta$$

which proves the lemma.

1.4. The geometry of Lie groups of polynomial volume growth. We present below the main families of Lie groups of polynomial volume growth and we give some results on their geometry and the behavior at infinity of the centered sub-Laplacians on these groups.

1. Simply connected nilpotent Lie groups. As we have already pointed out, all connected nilpotent Lie groups have polynomial volume growth.

A special class of simply connected nilpotent Lie groups are the stratified ones (cf. [**FS**]). These groups have a family of dilations. The dilation invariant sub-Laplacians are the easiest ones to study.

Every simply connected nilpotent Lie group N behaves at infinity as a stratified one N_0 (cf. [**NRS, V1**]). Furthermore, any centered left invariant sub-Lapacian L on N is converging at infinity to a dilation invariant sub-Laplacian L_0 on N_0.

2. Non-simply connected nilpotent Lie groups. Let N be a connected nilpotent Lie group and let us assume that it is not simply connected. Let C be the maximal torus of N. C is a compact central analytic subgroup of N and N/C is simply connected (cf. [**Ho** p. 188, **Va** pp. 195-200]). Let us denote by π the quotient map $\pi : N \to N/C$. Then, the limit group at infinity N_0 of N is the same as the limit group at infinity $(N/C)_0$ of N/C and the limit sub-Laplacian L_0 of L is the limit sub-Laplacian $d\pi(L)_0$ of its image $d\pi(L)$ on N/C.

3. The semidirect product $N \rtimes M$ of a simply connected nilpotent Lie group N by a compact Lie group M. As one might expect, the group $N \rtimes M$ has the same behavior at infinity as the group N. But the action of M on N gives rise to phenomena of homogenisation (cf. [**A1, A2**]). So the limit operator associated to a left invariant sub-Laplacian L on $N \rtimes M$ is called homogenised and denoted by L_H. It is a left invariant sub-Laplacian on N and is given by a formula similar to the formula that one has in the classical homogenisation theory (cf. [**BLP, JKO**]). This formula involves certain smooth functions ψ^j, which are called correctors and which are defined on M.

4. Simply connected solvable Lie groups. Let S be such a group (of polynomial volume growth) and let L be a centered left invariant sub-Laplacian on S. As all connected Lie groups, S has a largest normal analytic nilpotent subgroup $N \triangleleft S$, called nil-radical (cf. [**Va**]). We have that $[S,S] \subset N$. So, S/N is a simply connected abelian Lie group and hence it can be identified with $\mathbb{R}^k$.

We would like to view L as an operator on a nilpotent Lie group S_N and thus reduce its study to the nilpotent case.

The simplest case is when S splits into a semidirect product $S \cong N \rtimes \mathbb{R}^k$, with the action of $\mathbb{R}^k$ on N being semisimple. This is the case for example when S is algebraic. Then, as S_N we take the direct product $S_N = N \times \mathbb{R}^k$.

If S does not split into a semidirect product then the construction of S_N is more complicated. Intuitively, this is done as follows. We consider first a convenient section of S/N which we identify again with $\mathbb{R}^k$. The action of this section on N has a semisimple and a unipotent part. If we chose to ignore the semisimple part and keep only the unipotent part then we obtain a nilpotent group denoted by S_N.

We shall call S_N nil-shadow of S. This is done in analogy with Auslander and Green [**AG**], who used the term nil-shadow to describe the nilpotent part in the semisimple splitting of S. The semisimple splitting gives an imbedding of a connected solvable Lie group S into another solvable Lie group S' which is the semidect product $S' = N_S \rtimes A$ of a nilpotent Lie group N_S, the nil-shadow of S, by an abelian Lie group A (for the semisimple splitting and other splittings of S and their applications to the structure of solvmanifolds see [**A1, A2, Go**]).

It turns out that S_N is isomorphic to the nil-shadow constructed by Auslander and Green. This is trivial to verify when $S \cong N \rtimes \mathbb{R}^k$. In the general case, one has to go carefully through the proof of the semisimple splitting (the proof given in[**V3** pp. 404-405] is more appropriate for this purpose). The construction that we give here though of the nil-shadow S_N is more convenient for our purposes. It is also elementary, in the sense that it does not use the theory of algebraic groups.

Since S has polynomial volume growth, the eigenvalues of $\mathrm{Ad}\, x, x \in S$ are of the type e^{ia} with $a \in \mathbb{R}$. Thus the semisimple part of the action of $S/N \cong \mathbb{R}^k$ on N gives rise to rotations and hence to phenomena of homogenisation.

More precisely, let $*_S$ and $*_{S_N}$ denote the group products of S and S_N respectively. Let us also denote by $\mathfrak{s}$ and $\mathfrak{s}_N$ the Lie algebras of S and and S_N and identify them, respectively, with the $*_S$ and $*_{S_N}$-left invariant vector fields. Finally let us fix a base $\{X_1, ... X_q\}$ of $\mathfrak{s}_N$. The interest of S_N is due to the following fact:

Any $*_S$-left invariant vector field E can be written as

$$E(x) = a_1(x)X_1(x) + ... + a_q(x)X_q(x), \tag{1.4.1}$$

where the coefficients $a_1(x), ..., a_q(x)$ are quasi-periodic functions in the following sense:

Let us denote by π the quotient map $\pi : S \to \mathbb{R}^k \cong S/N$. Then for all $i = 1, ..., q$ there are constants $c_1, c_1', ..., c_j, c_j' \in \mathbb{R}$ and vectors $b_1, b_1' ..., b_j, b_j' \in \mathbb{R}^k$ such that

$$a_i(x) = \sum_{1 \le \ell \le j} c_\ell \sin \langle b_\ell, \pi(x) \rangle + c_\ell' \cos \langle b_\ell', \pi(x) \rangle .$$

So, any left invariant sub-Laplacian on S can be written as a differential operator with quasi-periodic coefficients on the nil-shadow S_N of S.

Again, as in the case of differential operators with periodic coefficients in $\mathbb{R}^n$ (cf. [**BLP**]), the macroscopic behavior of L is described by a homogenised operator L_H. L_H is a left invariant sub-Laplacian on S_N and it is given by a formula involving certain bounded smooth functions ψ^j, called correctors. In this case, the correctors are defined on $\mathbb{R}^k \cong S/N$ and are finite sums of periodic functions.

Let us denote by g^{-1_S} and $g^{-1_{S_N}}$ respectively the $*_S$ and $*_{S_N}$ inverses of g. Then, there is $c > 1$ such that for all $g, h \in S$

$$\frac{1}{c}|g^{-1_S} *_S h|_S \le |g^{-1_{S_N}} *_{S_N} h|_{S_N} \le c|g^{-1_S} *_S h|_S.$$

5. Non-simply connected solvable Lie groups. Let S be a connected solvable Lie group of polynomial volume growth and let us assume that it is not simply connected. As in section 1.4.2, let us consider the maximal torus C of the nilradical N of S. Then N/C is simply connected. Since C is fully invariant in N it is also normal in S. So we can consider the group S/C. Since the group $(S/C)/(N/C) \cong S/N$ is abelian, we have that $(S/C)/(N/C) \cong \mathbb{R}^k \times \mathbb{T}^m$ (cf. [**Ho** p. 39]), where $\mathbb{T} = \mathbb{R}/\mathbb{Z}$. Let π be the projection $\pi : S/C \to \mathbb{R}^k \times \mathbb{T}^m$ and let $S_1 = \pi^{-1}(\mathbb{R}^k)$. Then S_1 is simply connected and solvable and $(S/C)/S_1 \cong \mathbb{T}^m$ is compact. Hence S_1 is a semidirect factor in S/C, i.e. $S/C \cong S_1 \rtimes \mathbb{T}^m$ (cf. [**Ho** pp. 139-140, **Va** pp. 254-256]) .

Let us denote by π the quotient map $\pi : S \to S/C$. Then, the homogenised sub-Laplacian L_H of a left invariant sub-Laplacian L on S is the same as the homogenised sub-Laplacian $d\pi(L)_H$ of its image $d\pi(L)$ on S/C, i.e. it is a left invariant sub-Laplacian on the nil-shadow of $(S_1)_N$ of S_1. The correctors ψ^j will still be finite sums of periodic functions defined this time on $\mathbb{R}^k \times \mathbb{T}^m \cong S/N$.

6. Simply connected Lie groups of polynomial volume growth. Every connected Lie group G admits a Levi decomposition (cf. [**Va**]), i.e. it has a largest solvable normal analytic subgroup $S \lhd G$ called the radical of G and a maximal semisimple analytic subgroup M, called Levi subgroup, such that $G = SM$. M is not necessarily unique. G also has a largest nilpolent normal analytic subgroup $N \lhd G$, called the nil-radical of G. Both S and N are closed in G. Also $N \lhd S$ and $[G, G] \subseteq NM$.

If G has polynomial volume growth then M is compact.

If G is simply connected, then $S \cap M = \{e\}$ and hence $G \cong S \rtimes M$. Also $G/N \cong \mathbb{R}^k \times M$.

Now, let us assume that G is a simply connected Lie group of polynomial volume growth. Then the homogenised sub-Laplacian L_H of a left invariant sub-Laplacian L on G is a left invariant sub-Laplacian on the nil-shadow S_N of S. The correctors ψ^j will be finite sums of bounded smooth functions $\phi(x, z)$ which are defined on $\mathbb{R}^k \times M$ and which are periodic with respect to the variable $x \in \mathbb{R}^x$.

7. Non-simply connected Lie groups of polynomial volume growth. Let G be a connected Lie group of polynomial volume growth and let us assume that G is not simply connected. Let $G = SM$ a Levi decomposition of G. Since G is not simply connected, it is not necessarily true that $S \cap M = \{e\}$. To get around this difficulty, we proceed as in section 1.4.5 and we consider the maximal torus C of N. Again, since C is fully invariant in N, is also a normal subgroup of G.

Let us denote by π the quotient map $\pi : G \to G/C$. As in section 1.4.5 we can see that the radical S/C of G/C is isomorphic to the semidirect product $S/C \cong S_1 \rtimes T$ of a simply connected solvable subgroup $S_1 \lhd S/C$ and a compact abelian analytic subgroup T of G/C.

We set $M_1 = T\ (MC/C)$. Then M_1 is a compact subgroup of G/C and $G/C = S_1 M_1$. Since S_1 is simply connected, we also have $S_1 \cap M_1 = e$ (cf. [**Ho** p. 138]) and hence $G/C \cong S_1 \rtimes M_1$.

Note also that $G/N \cong (G/C)/(N/C) \simeq \mathbb{R}^k \rtimes M_1$.

Let us denote by $*_{(S_1)_N}$ the group product of $(S_1)_N$ and by $x^{-1_{(S_1)_N}}$ the $*_{(S_1)_N}$-inverse of $x \in (S_1)_N$. Then, there is $c > 1$ such that for all $g, h \in G$ with $\pi(g) = (x, z), \pi(h) = (y, w) \in S_1 \rtimes M_1$

$$\frac{1}{c}|g^{-1}h|_G \leq |x^{-1_{(S_1)_N}} *_{(S_1)_N} y|_{(S_1)_N} \leq c|g^{-1}h|_G.$$

The homogenised sub-Laplacian L_H of a left invariant sub-Laplacian L on G is the same as the homogenised sub-Laplacian $d\pi(L)_H$ of its image $d\pi(L)$ on G/C, i.e. L_H is a left invariant sub-Laplacian on the nil-shadow $(S_1)_N$ of S_1. The correctors ψ^j will be finite sums of bounded smooth functions $\phi(x, z)$ which are defined on $\mathbb{R}^k \times M_1$ and which are periodic with respect to the variable $x \in \mathbb{R}^k$.

Since $C \lhd G$ there are Haar measures $d_{G/C}g$ and $d_C g$ on G/C and C respectively so that the Haar measure dg on G desintegrates as follows (cf. [**Bou1**]):

$$\int_G f(g)dg = \int_{G/C} \int_C f(gz) d_C z \, d_{G/C}\pi(g).$$

Since C is compact we can and shall assume that d_C-measure$(C) = 1$.

Also since $G/C \cong S_1 \rightthreetimes M_1$ there are Haar measures $d_{S_1}g$ and $d_{M_1}g$ on S_1 and M_1 respectively such that

$$\int_{G/C} f(g)dg = \int_{S_1}\int_{M_1} f(x,z)d_{S_1}xd_{M_1}z.$$

Since M_1 is compact we shall also assume that d_{M_1}-measure$(M_1) = 1$.

We shall drop the indices G/C, C, S_1 and M_1 when there is no risk of confusion.

Motivated by the specific nature of the correctors we shall give the following definition:

We shall say that a function f on $\mathbb{R}^k \times M_1$ is of type P if there is $a \in \mathbb{R}^k$ and a function $\phi \in C^\infty(M_1)$ such that either

$$f(x,z) = \sin(\langle a, x\rangle))\phi(z), \qquad (x,z) \in \mathbb{R}^k \times M_1,$$

or

$$f(x,z) = \cos(\langle a, x\rangle))\phi(z), \qquad (x,z) \in \mathbb{R}^k \times M_1.$$

Let us denote by π_0 the canonical projection $\pi_0 : G \to \mathbb{R}^k \times M_1$.

We shall say that f is a function of type P on G if there is another function f' of type P on $\mathbb{R}^k \times M_1$ such that $f = f' \circ \pi_0$.

We shall say that f is a function of type QP on G if it is a finite sum of functions of type P.

If f is a function of type QP then we shall denote by $\langle f\rangle$ its mean value

$$\langle f\rangle = \lim_{n\to\infty} \frac{1}{|U^n|}\int_{U^n} f(x)dx$$

where U is a compact neighborhood of the identity element e of G.

Note that if f is a function of type QP then Lf is also a function of type QP. If we also have $\langle f\rangle = 0$, then there is a unique function u of type QP satisfying

$$Lu = f, \qquad \langle f\rangle = 0.$$

The correctors ψ^j will be functions of type QP satisfying $\langle \psi^j\rangle = 0$.

1.5. Calculation of the homogenised sub-Laplacian in the case of the group $\mathbb{R}^q \rightthreetimes M$. In order to illustrate what was said in the previous section, we shall calculate the homogenised sub-laplacian in the case of the semidirect product $\mathbb{R}^q \rightthreetimes M$ of $\mathbb{R}^q$ by a connected compact Lie group M.

Let us denote by ϑ_z, $z \in M$ the action of M on $\mathbb{R}^q$. Note that if $(x,z),(y,w) \in \mathbb{R}^q \rightthreetimes M$ then

$$(x,z)(y,w) = (x + \vartheta_z(y), zw).$$

Let us denote by ϑ_{zij} the entries of the matrix of ϑ_z wih respect to the canonical basis $\{e_1, ..., e_q\}$ of $\mathbb{R}^q$. If X is a left invariant vector field on $\mathbb{R}^q \rightthreetimes M$ satisfying $X(e) = \frac{\partial}{\partial x_j}$, then

$$X(x,z) = \sum_{1\leq j\leq q} \vartheta_{zij}\frac{\partial}{\partial x_i} \qquad (1.5.1)$$

By writing every $v \in \mathbb{R}^q$ as

$$v = \int_M \vartheta_z(v)dz + \left(v - \int_M \vartheta_z(v)dz\right)$$

we have the decomposition

$$\mathbb{R}^q = V_0 \oplus V_1$$

with

$$V_0 = \{v \in \mathbb{R}^q : \vartheta_z(v) = v,\ z \in M\} \quad \text{and} \quad V_1 = \{v \in \mathbb{R}^q : \int_M \vartheta_z(v)dz = 0\}.$$

Note that with the notation of section 1.2, we have $H = V_1 \rtimes M$.

By chosing a different basis of $\mathbb{R}^q$ is necessary, we can assume that $\{e_1, ..., e_n\}$ is a basis of V_0 and $\{e_{n+1}, ..., e_q\}$ a basis of V_1. Then

1. if either $1 \leq i \leq n$ or $1 \leq j \leq n$, then $\vartheta_{zij} = \delta_{ij}$, where $\delta_{ij} = 1$ for $i = j$ and $\delta_{ij} = 0$ for $i \neq j$ and
2. if $n < i, j \leq q$, then $\langle \vartheta_{zij} \rangle = 0$.

1. L written as an operator on $\mathbb{R}^q \times M$. Let us fix a basis $\{Z_1, ..., Z_m\}$ of the Lie algebra $\mathfrak{m}$ of M.

Let $L = -(E_1^2 + ... + E_p^2) + E_0$ be a centered left invariant sub-Laplacian on $\mathbb{R}^q \rtimes M$. Then L can be written as

$$(1.5.2) \quad L = -\sum_{1 \leq i,j \leq m} Z_i a_{ij} Z_j - \sum_{\substack{1 \leq i \leq m \\ 1 \leq j \leq q}} Z_i b_{ij} \frac{\partial}{\partial x_j} - \sum_{\substack{1 \leq i \leq q \\ 1 \leq j \leq m}} \frac{\partial}{\partial x_j} b_{ji} Z_i - \sum_{1 \leq i,j \leq q} \frac{\partial}{\partial x_i} \zeta_{ij} \frac{\partial}{\partial x_j} + \sum_{1 \leq i \leq m} a_i Z_i + \sum_{1 \leq i \leq q} \zeta_i \frac{\partial}{\partial x_i}$$

where the coefficients are analytic functions defined on M and satisfying:

1. a_{ij} =const., $1 \leq i, j \leq m$,
2. ζ_{ij} =const., $1 \leq i, j \leq n$,
3. if $1 \leq j \leq n$, then b_{ij} =const.
4. for all $1 \leq i \leq m$, a_i =const.
5. if $1 \leq i \leq n$, then $\zeta_i = 0$ and finally,
6. if $n < i \leq q$, $\langle \zeta_i \rangle = 0$.

2. The correctors and the homogenised operator L_H. Let us fix a function $f \in C^\infty(\mathbb{R}^q)$ and let us extend f to $\mathbb{R}^q \times M$ by setting $f(x, z) = f(x)$, $(x, z) \in N \times M$. Then, by (1.5.2),

$$(1.5.3) \qquad Lf = -\sum_{1 \leq i,j \leq q} \zeta_{ij} \frac{\partial}{\partial x_i} \frac{\partial}{\partial x_j} f - \sum_{\substack{1 \leq i \leq m \\ n < j \leq q}} (Z_i b_{ij}) \frac{\partial}{\partial x_j} f + \sum_{n < i \leq q} \zeta_i \frac{\partial}{\partial x_i} f.$$

The definition of the correctors ψ^j, $j = 1, ..., q$ is motivated by this remark. Let us first denote by L_M the projection of L on M:

$$L_M = -\sum_{1 \leq i,j \leq m} Z_i a_{ij} Z_j + \sum_{1 \leq i \leq m} a_i Z_i.$$

DEFINITION. We define the (first order) correctors $\psi^j, n < j \leq q$ (cf. [**BLP, JKO**]) as solutions of the problem

$$L_M \psi^j = -\zeta_j + \sum_{1\leq i\leq m} Z_i b_{ij}, \quad \langle \psi^j \rangle = 0. \tag{1.5.4}$$

If $1 \leq j \leq n$ then we just set $\psi^j = 0$.

Note that $\langle Z_i b_{ij} \rangle = 0$, $1 \leq i \leq m$ and that by the property (6) in section 1.5.1, $\langle \zeta_j \rangle = 0$, $n < j \leq q$. So the correctors are well defined.

Combining (1.5.3) and (1.5.4), we have that there is $c > 0$ independent of f such that

$$\begin{aligned} &L\left(f + \sum_{1\leq j\leq q} \psi^j \frac{\partial}{\partial x_j} f\right) \\ &= -\sum_{1\leq i,j\leq q} \left(\zeta_{ij} - \zeta_i \psi^j + \sum_{1\leq \ell\leq m} b_{\ell i} Z_\ell \psi^j + \sum_{1\leq \ell\leq m} Z_\ell \left(b_{\ell i}\psi^j\right)\right) \frac{\partial}{\partial x_i}\frac{\partial}{\partial x_j} f + F \end{aligned} \tag{1.5.5}$$

with the function F satisfying $|F(x,z)| \leq c|\nabla^3 f(x)|$, $(x,z) \in \mathbb{R}^q \times M$.

This expression motivates the following definitions:

DEFINITION. The homogenised sub-Laplacian L_H associated to L is defined to be the operator

$$L_H = -\sum_{1\leq i,j\leq q} q_{ij} \frac{\partial}{\partial x_i}\frac{\partial}{\partial x_j}$$

with coefficients defined by

$$q_{ij} = \left\langle \zeta_{ij} - \zeta_i \psi^j + \sum_{0\leq \ell\leq m} b_{\ell i} Z_\ell \psi^j \right\rangle, \qquad 1 \leq i,j < q.$$

DEFINITION. We define the (second order) correctors $\psi^{ij}, 1 \leq i,j \leq q$ (cf. [**BLP, JKO**]) as solutions of the problem

$$L_M \psi^{ij} = \zeta_{ij} - \zeta_i \psi^j + \sum_{1\leq \ell\leq m} b_{\ell i} X_\ell \psi^j + \sum_{1\leq \ell\leq m} Z_\ell \left(b_{\ell i}\psi^j\right) - q_{ij}, \ \ \langle \psi^{ij} \rangle = 0.$$

Combining the above defionitions with (1.5.5) we have the following expression which explains the relation between the operators L and L_H:

$$L\left(f + \sum_{1\leq j\leq n_2} \psi^j \frac{\partial}{\partial x_j} f + \sum_{1\leq i,j\leq q} \psi^{ij} \frac{\partial}{\partial x_i}\frac{\partial}{\partial x_j} f\right) = L_H f + F \tag{1.5.6}$$

with the function F satisfying $|F(x,z)| \leq c\left(|\nabla^3 f(x)| + |\nabla^4 f(x)|\right), (x,z) \in N \times M$.

3. L_H is an elliptic operator on $\mathbb{R}^q$. To prove this, it is enough to prove that if $\xi = (\xi_1, ..., \xi_{n_1}) \in \mathbb{R}^{n_1}, \xi \neq 0$, then

$$\sum_{1\leq i,j\leq q} q_{ij}\xi_i\xi_j > 0. \tag{1.5.7}$$

To this end, let us consider the function

$$u(x,z)=\sum_{1\le i\le q}\xi_i\left(x_i+\psi^i(z)\right),\quad (x,z)\in\mathbb{R}^q\times M.$$

Then, by (1.5.4), $Lu=0$. Furthermore

$$Lu^2=2uLu-2[(E_1u)^2+...+(E_pu)^2]=-2[(E_1u)^2+...+(E_pu)^2]\le 0. \tag{1.5.8}$$

We want to prove that

$$\left\langle Lu^2\right\rangle=-2\sum_{1\le i,j\le q}q_{ij}\xi^i\xi^j. \tag{1.5.9}$$

We have

$$\begin{aligned}u^2&=\sum_{1\le i,j\le q}\xi_i\xi_j(x_i+\psi^i)(x_j+\psi^j)\\&=\sum_{1\le i,j\le q}\xi_i\xi_j(x_ix_j+x_i\psi^j+x_j\psi^i+\psi^i\psi^j).\end{aligned}$$

Also

$$\begin{aligned}L\left[(x_i+\psi^i)(x_j+\psi^j)\right]=&-\zeta_{ij}-\zeta_{ji}+\zeta_i\psi^j+\zeta_j\psi^i+L(\psi^i\psi^j)\\&-\sum_{1\le\ell\le m}\left[Z_\ell(b_{\ell i}\psi^j)+b_{\ell i}Z_\ell\psi^j+Z_\ell(b_{\ell j}\psi^i)+b_{\ell j}Z_\ell\psi^i\right].\end{aligned}$$

Since $\left\langle L(\psi^i\psi^j)\right\rangle=0$, we conclude that

$$\left\langle L(\left(x_i+\psi^i\right)\left(x_j+\psi^j\right))\right\rangle=-2q_{ij}$$

which proves (1.5.9).

Now, if we had

$$\sum_{1\le i,j\le q}q_{ij}\xi_i\xi_j=0$$

then by (1.5.8), we would have that $E_iu=0$, for all $1\le i\le p$, i.e. the function u would be constant along the integral curves of the vector fields $E_1,...,E_p$. Since these vector fields satisfy Hörmander's condition, u would then be a constant function, and hence we would have that

$$\xi^1x_1+...+\xi^qx_q=\xi^1\psi^1+...+\xi^q\psi^q+\text{const}.$$

This is absurd because the right term is a bounded function and the left term is not bounded.

6. A parabolic Harnack Inequality.

THEOREM 1.6.1. *Let L be a centered left invariant sub-Laplacian on a connected Lie group of polynomial volume growth G. Let also U a compact neighborhood of the identity element e of G and let $a,b\in\mathbb{N}$. Then there are $\alpha,\beta,\theta\in\mathbb{N},\alpha<\beta$ and $c>0$ such that for all $n\in\mathbb{N}$ and all $u\ge 0$ satisfying*

$$\left(\frac{\partial}{\partial t}+L\right)u=0\quad in\quad (0,\theta n^2)\times U^{\theta n},$$

we have

$$\sup\left\{u(\alpha n^2,x):x\in U^{an}\right\}\le c\inf\left\{u(\beta n^2,x):x\in U^{bn}\right\}. \tag{1.6.1}$$

When the drift term $E_0 = 0$, then the elliptic version of the above inequality was proved by Varopoulos [**V2**] by adapting the method of Moser [**M1**]. The parabolic version, also for sub-Laplacians with drift $E_0 = 0$, was proved by Salof-Coste [**SC**] by using a Gaussian estimate on the spacial derivatives of the heat kernel.

1. The strategy of the proof. This inequality is proved by the method of Krylov and Safonov [**KS**]. This method uses certain information on the growth of the positive heat functions. To obtain this information we use the following three results.

The first result is a theorem of Varopoulos [**V6**], which asserts that the heat kernel $p_t(x, y)$ of L decreases with a certain uniform speed as $t \to \infty$:

THEOREM 1.6.2 (Varopoulos [**V6**]). *Let L be a non-necessarily centered left invariant sub-Laplacian on G and let $p_t(x, y)$ the heat kernel of L. Then there is a constant $c > 0$ such that*

$$p_t(x, y) \leq c\, t^{-D/2}, \quad t \geq 1,\ x, y \in G. \tag{1.6.2}$$

The other two results concern the distribution of the mass of the heat kernel $p_t(x, y)$ as $t \to \infty$:

PROPOSITION 1.6.3. *Let $p_t(x, y)$ be the heat kernel of a centered left invariant sub-Laplacian L on a connected Lie group of polynomial volume growth G and let U be a compact neighborhood of the identity element e of G. Then for all $a \in \mathbb{N}$ there is $\delta > 0$ such that*

$$\int_{U^n} p_t(x, y)dy > \delta, \tag{1.6.3}$$

for all $n \in \mathbb{N}$, $a^{-2}n^2 \leq t \leq a^2n^2$ and $x \in U^{an}$.

PROPOSITION 1.6.4. *Let L, $p_t(x, y)$ and U be as above. Then for all $\delta > 0$ there is $a \in \mathbb{N}$ such that*

$$\int_{y \notin U^{an}} p_t(e, y)dy < \delta, \tag{1.6.4}$$

for all $n \in \mathbb{N}$, $0 < t \leq n^2$.

The above propositions are rather obvious in the case of dilation invariant sub-Laplacians on stratified nilpotent groups. In the case of general connected nilpotent Lie groups, they are proved by making use of the associated limit sub-Laplacian. In the case of more general connected Lie groups of polynomial volume growth they are proved by making use of the associated homogenised sub-Laplacian.

2. The oscillation of the heat functions. If $B \subseteq \mathbb{R} \times G$ and u is a function defined on B, then let us set

$$\mathrm{Osc}(u, B) = \sup\{|u(t, x) - u(s, y)| : (t, x), (s, y) \in B\}.$$

In the course of the proof of (1.6.1) we shall prove the following result:

PROPOSITION 1.6.5. *There is $c \in \mathbb{N}$ and $\alpha \in (0, 1)$ such that*

$$\mathrm{Osc}\left(u, (t - n^2, t) \times U^n\right) \leq \alpha\, \mathrm{Osc}\left(u, (t - c^2n^2, t) \times U^{cn}\right) \tag{1.6.5}$$

for all $n \in N$ and all functions u satisfying

$$\left(\frac{\partial}{\partial t} + L\right) u = 0 \quad \textit{in} \quad (t - c^2n^2, t) \times U^{cn}.$$

1.7. Positive harmonic functions. A consequence of the Harnack inequality (1.6.1) is the following result:

THEOREM 1.7.1. *Let L be a centered left invariant sub-Laplacian on a connected Lie group G of polynomial volume growth. Then, every positive function $u \geq 0$ satisfying $Lu = 0$ on G is constant.*

1.8. Gaussian estimates for the heat kernel. Making use of the Harnack inequality (1.6.1) we can prove the following upper Gaussian estimate:

THEOREM 1.8.1. *Let $p_t(x, y)$ be the heat kernel of a centered left invariant sub-Laplacian L on a connected Lie group G of polynomial volume growth and let $|.|_G$ be as in (1.1). Then there is a constant $c > 0$ such that*

$$p_t(x, y) \leq c\, t^{-D/2} \exp\left(-\frac{|x^{-1}y|_G^2}{ct}\right), \quad t \geq 1,\ x, y \in G. \tag{1.8.1}$$

When $E_0 \in \mathfrak{e}_1 + \mathfrak{e}_2$ then we also have a small time (i.e. valid for $0 < t \leq 1$) Gaussian estimates (cf. [**He, V4, V7** appendix A.4]). If $E_0 \notin \mathfrak{e}_1 + \mathfrak{e}_2$ then the situation is much more complicated (cf. [**BL1, BL2**]).

We also have the following lower Gaussian estimate:

COROLLARY 1.8.2. *Let L and $p_t(x, y)$ be as above. Then there is a constant $c \geq 1$ such that*

$$p_t(x, y) \geq \frac{1}{c}\, t^{-D/2} \exp\left(-c\frac{|x^{-1}y|_G^2}{t}\right) \tag{1.8.2}$$

for all $t \geq 1$, and $x, y \in G$.

For time $t = 1$, the above estimate has been proved by Varopoulos (cf. [**V4, V5, V8**]). If $|x^{-1}y|_G \geq t/c$, then (1.8.2) follows be repeating the lower estimate for $t = 1$. If $|x^{-1}y|_G \leq t/c$, then (1.8.2) follows from the Harnack inequality (1.6.1) by arguing as in [**VSC,** pp. 47-50].

1.9. A Taylor formula for the heat functions on nilpotent Lie groups

1. Polynomials on nilpotent Lie groups. If N is a simply connected nilpotent Lie group, then using the exponential coordinates we can identify N, as a differential manifold, with $\mathbb{R}^q$. So, by a monomial on N we shall understand a monomial on $\mathbb{R}^q$.

For every monomial $P(x)$, there is a unique integer $k \geq 0$ such that for every compact neighborhood U of e there is $c > 0$ such that

$$\frac{1}{c} n^k \leq \sup\{|P(x)|,\ x \in U^n\} \leq cn^k, \quad n \in \mathbb{N}. \tag{1.9.1}$$

We shall define the homogeneous degree $\deg_H P$ of $P(x)$ by $\deg_H P = k$.

We shall say that $P(t, x)$ is a monomial on $\mathbb{R} \times N$ if $P(t, x) = t^m Q(x)$, with $Q(x)$ a monomial on N. We shall define the homogeneous degree $\deg_H P(t, x)$ of $P(t, x)$ by

$$\deg_H P(t, x) = 2m + \deg_H Q(x).$$

By polynomials we shall of course understand linear combinations of monomials. The homogeneous degree of a polynomial is therefore the maximum of the homogeneous degrees of its monomials.

PROPOSITION 1.9.1. *Let L be a centered left invariant sub-Laplacian on N and let L_0 be the associated limit operator. Then to any monomial $P(t,x)$ we can associate a polynomial $P(t,x)$ satisfying*

$$\begin{gathered} Q_P(t,x) = P(t,x) + W(t,x) \\ \deg_H W \leq \deg_H P - 1 \\ \left(\frac{\partial}{\partial t} + L_0\right) P(t,x) = \left(\frac{\partial}{\partial t} + L\right) Q_P(t,x). \end{gathered} \tag{1.9.2}$$

Note that the polynomial $Q_P(t,x)$ in the above proposition, is not necessarily unique.

From now on, we shall denote, for all $k \in \mathbb{N}$, by

$$P_0(t,x), P_2(t,x), ..., P_{\nu_k}(t,x)$$

the monomials with homogeneous degree $\leq k$ and we shall fix, once and for all, polynomials

$$Q_{P_0}(t,x), Q_{P_2}(t,x), ..., Q_{P_{\nu_k}}(t,x)$$

satisfying (1.9.2).

2. A Taylor formula for the heat functions.

The following result gives a Taylor formula for the heat functions. The proof is based on ideas from [**AL1, Al2**]. These ideas have already been used in the context of Lie groups in [**A1, A2, ALo**]. The interest of the method lies in the fact that we do not make use of any a priori control on the derivatives of the heat functions.

THEOREM 1.9.2. *For all $k \in \mathbb{N}$ there is $c_k > 0$ such that for all $n \in \mathbb{N}$ and all functions u satisfying*

$$\left(\frac{\partial}{\partial t} + L\right) u = 0, \quad \text{in } (-n^2, n^2) \times U^n$$

we have

$$\sup\left\{|u - \sum_{0\leq i\leq \nu_k} A_i n^{-\deg_H P_i} Q_{P_i}|; (-1,1)\times U\right\} < c_k n^{-(k+1)}\|u\|_\infty \tag{1.9.3}$$

where the constants A_i satisfy

$$|A_i| \leq c_k \|u\|_\infty, \qquad 0 \leq i \leq \nu_k$$

and where

$$\left(\frac{\partial}{\partial t} + L\right)\left(\sum_{\nu_{j-1}<i\leq\nu_j} A_i Q_{P_i}\right) = 0, \qquad 1 \leq j \leq k.$$

1.10. Harmonic functions of polynomial growth on connected nilpotent Lie groups. Let G be a connected Lie group and let U be a compact neighborhood of the identity element e of G. We shall say that a function u on G grows polynomially, if there is $c > 0$ and $A > 0$ such that

$$\sup\{|u|; U^n\} \leq cn^A, \quad n \in \mathbb{N}. \tag{1.10.1}$$

An application of theorem 1.9.2 is the following:

THEOREM 1.10.1. *Let L be a centered left invariant sub-Laplacian on a connected nilpotent Lie group N. Then every L-harmonic function u on N (i.e. satisfying $Lu = 0$ on N), which grows polynomially, is equal to a polynomial.*

We shall give the generalisation of the above result to connected Lie groups of polynomial volume growth in section 1.12 below, where we shall also make some further remarks and give some related references.

1.11. A Taylor formula for the heat functions on Lie groups of polynomial volume growth.

1. The corrected monomials. We shall use the notation of sections 1.4 and 1.9. Let G be a connected Lie group of polynomial volume growth. We define monomials on G by extending the monomials $P(x)$ on $(S_1)_N$ in the following way:
We extend first the monomials $P(x)$ on $(S_1)_N$ to monomials $P'(x,z)$ on $S_1 \leftthreetimes M_1 \cong G/C$ by setting

$$P'(x,z) = P(x), \qquad (x,z) \in S_1 \leftthreetimes M_1$$

and then to monomials $P''(g)$ on G by setting

$$P''(g) = P'(\pi(g)), \qquad g \in G.$$

If the original monomial P on $(S_1)_N$ satisfies (1.9.1) then its extention P'' to G also satisfies (1.9.1) i.e. P and its extention P'' to G have the same homogeneous degree.

As in the nilpotent case, we shall say that $P(t,g)$ is a monomial on $\mathbb{R} \times G$ if $P(t,g) = t^m Q(g)$, with $Q(g)$ a monomial on G. We shall define the homogeneous degree
$\deg_H P(t,g)$ of $P(t,g)$ by

$$\deg_H P(t,g) = 2m + \deg_H Q(g).$$

As in the nilpotent case, we shall denote by

$$P_0(t,g), P_2(t,g), ..., P_{\nu_k}(t,g)$$

the monomials with homogeneous degree $\leq k$.

Let L be a centered left invariant sub-Laplacian on G and let L_H be the associated homogenised operator.

Let us also associate to L_H and to the monomials $P_i(t,x)$, $i = 0,1,2,...$ polynomials $Q_{P_i}(t,x)$, $i = 0,1,2,...$, as in section 1.9.1.

PROPOSITION 1.11.1. *To every monomial $P_i(t,g)$ with $\deg_H P_i = k$, as above, we can associate another "corrected monomial" $Q^{\psi}_{P_i}(t,g)$ which can be written as*

$$Q^{\psi}_{P_i}(t,g) = P_i(t,g) + \sum_{0 \leq j \leq \nu_{k-1}} \psi^i_j(g) P_j(t,g), \tag{1.11.1}$$

where the functions ψ^i_j are functions of type QP and which satisfies

$$\left(\frac{\partial}{\partial t} + L_H\right) Q_{P_i}(t,g) = \left(\frac{\partial}{\partial t} + L\right) Q^{\psi}_{P_i}(t,g). \tag{1.11.2}$$

Note that the polynomials $Q^{\psi}_{P_i}$ in (1.11.1) and (1.11.2) above, are not necessarily unique.

In the rest of this article, to every monomial P_i, $i \in \mathbb{N}$, we shall associate and fix, once and for all, a corrected monomial $Q^{\psi}_{P_i}$, satisfying (1.11.1) and (1.11.2).

2. A Taylor formula for the heat functions. The following result is the generalisation of theorem 1.9.2.

THEOREM 1.11.2. *For all $k \in \mathbb{N}$ there is $c_k > 0$ such that for all $n \in \mathbb{N}$ and all functions u satisfying*

$$\left(\frac{\partial}{\partial t} + L\right) u = 0, \quad in\ (-n^2, n^2) \times U^n$$

we have

$$\sup\left\{ |u - \sum_{0 \le i \le \nu_k} A_i n^{-\deg_H P_i} Q^{\psi}_{P_i}|; (-1,1) \times U \right\} < c_k n^{-(k+1)} \|u\|_\infty, \tag{1.11.3}$$

where the constants A_i satisfy

$$|A_i| \le c_k \|u\|_\infty$$

for all $0 \le i \le \nu_k$ and where

$$\left(\frac{\partial}{\partial t} + L\right)\left(\sum_{\nu_{j-1} < i \le \nu_j} A_i Q^{\psi}_{P_i}\right) = 0, \qquad 1 \le j \le k.$$

1.12. Harmonic functions of polynomial growth. The following result is an application of theorem 1.11.2 and generalises theorem 1.10.1.

THEOREM 1.12.1. *Let L be a centered left invariant sub-Laplacian on a connected Lie group G of polynomial volume growth. Then every L-harmonic function u on G (i.e. satisfying $Lu = 0$ on G) which grows polynomially, is equal to a linear combination of corrected monomials $Q^{\psi}_{P_i}$.*

A result of this type has first been proved by Avellaneda and Lin [**AL2**] in the case of differential operators with periodic coefficients in $\mathbb{R}^n$. A generalisation of this result for symmetric sub-Laplacians on Lie groups of polynomial volume growth was given in [**ALo**], where it was also used to prove a Sobolev inequality.

Finally, we point out that in the context of Riemmannian manbifolds with non-negative Ricci curvature, there is a wellknown conjecture of S.T.Yau which says that the linear space of harmonic functions with polynomial growth of a fixed rate is finite dimensional. For results and further references related to this conjecture we refer the reader to the recent paper of Colding and Minicozzi II [**CM**].

1.13. Harnack inequalities for the derivatives of the heat functions.

1. Nilpotent Lie groups. An consequence of theorem 1.9.2 is the following:

THEOREM 1.13.1. *Let L be a centered left invariant sub-Laplacian on a simply connected nilpotent Lie group N. Let also U be a compact neighborhood of the identity element e of N and let $a, b \in \mathbb{N}$. Then, for all $k \in \mathbb{N}$ and all left invariant* *vector fields* $Y_1, ..., Y_m$ *there are* $\alpha, \beta, \theta \in \mathbb{N}, \alpha < \beta < \theta$ *and* $c > 0$ *such that for all* $n \in \mathbb{N}$ *and all* $u \geq 0$ *satisfying* $\left(\frac{\partial}{\partial t} + L\right) u = 0$ *in* $(0, \theta n^2) \times U^{\theta n}$,

$$(1.13.1) \quad \sup\left\{|\frac{\partial^k}{\partial t^k} Y_1 Y_2 ... Y_m u(\alpha n^2, x)| : x \in U^{an}\right\} \leq cn^{-2k-m} \inf\left\{u(\beta n^2, x) : x \in U^{bn}\right\}.$$

An alternative way of proving the above result is by adapting the local theory in [**V3, VSC**] using the family of dilations associated with the limit group at infinity (cf. [**NRS**]).

2. Lie groups of polynomial volume growth. Unfortunately, the Harnack inequality (1.13.1) is not necessarily true for $m \geq 2$ when the group is not nilpotent.

To see this let us consider the group $\mathbb{R}^q \rtimes M$ studied in section 1.5 and let us consider the function

$$u(x, z) = x_j + \psi^j(z), \quad (x, z) \in \mathbb{R}^q \rtimes M$$

where $n < j \leq q$. This function grows linearly, i.e. there is $c > 0$ such that

$$\sup\{|u| \; ; U^r)\} \leq cr, \quad r \geq 1$$

and it satisfies $Lu = 0$. Also, $Z_i u = Z_i \psi^j$, $1 \leq i \leq m$. So, if $Z_i Z_j \psi^j \neq 0$ for some $1 \leq i, j \leq m$ and U is chosen large enough, so that $0 \times M \subseteq U$, then the inequality

$$\sup\{|Z_i Z_j u| \; ; \; U\} \leq cr^{-2} \sup\{|u| \; ; \; U^r\}, \quad r \geq 1$$

is false.

THEOREM 1.13.2. *Let* L *be a centered left invariant sub-Laplacian on a connected Lie group* G *of polynomial volume growth. Let also* U *be a compact neighborhood of the identity element* e *of* G *and* $a, b \in \mathbb{N}$. *Then, for all* $k \in \mathbb{N}$ *and all left invariant vector fields* Y *there are* $\alpha, \beta, \theta \in \mathbb{N}, \alpha < \beta < \theta$ *and* $c > 0$ *such that for all* $n \in \mathbb{N}$ *and all* $u \geq 0$ *satisfying* $\left(\frac{\partial}{\partial t} + L\right) u = 0$ *in* $(0, \theta n^2) \times U^{\theta n}$,

$$(1.13.2) \quad \sup\left\{|\frac{\partial^k}{\partial t^k} Y u(\alpha n^2, x)| : x \in U^{an}\right\} \leq cn^{-2k-1} \inf\left\{u(\beta n^2, x) : x \in U^{bn}\right\}.$$

Nevertherless, we can still controll the higher order spatial derivatives by differentiating the expression in the left part of (1.11.3). The expression that we shall obtain though, will involve other spacial derivatives of lower order as well as the correctors ψ_j^i and their derivatives (see for example theorem 1.14.9 below).

1.14. Berry-Esseen estimates.

1. Nilpotent Lie groups. Let L be a centered left invariant sub-Laplacian on a simply connected nilpotent Lie group N and let L_0 be the associated limit operator. Let also $p_t(x, y)$ and $p_t^0(x, y)$ be the heat kernels of L and L_0 respectively.

We have the following Berry-Esseen type of estimate (cf. [**Fe, Pe**]):

THEOREM 1.14.1. *There are constants* $c, C_L \geq 0$ *such that*

$$(1.14.1) \qquad |p_t(x, e) - p_t^0(x, e)| \leq ct^{-(D+1)/2}, \qquad x \in N, \; t \geq 1.$$

Since L_0, hence also $p_t^0(x, y)$, is dilation invariant, there is a constant $C_L > 0$ such that

$$(1.14.2) \qquad p_t^0(e, e) = c_L t^{-D/2}, \qquad t > 0.$$

Combining (1.14.1) and (1.14.2) we have the following:

COROLLARY 1.14.2. *There are constants* $c, C_L \geq 0$ *such that*

$$|p_t(e,e) - C_L t^{-D/2}| \leq ct^{-(D+1)/2}, \qquad t \geq 1. \tag{1.14.3}$$

Also by interpolating (1.8.1) and (1.14.1) we have the following:

COROLLARY 1.14.3. *For all* $\varepsilon \in (0,1)$ *there is a constant* $c > 0$ *such that*

$$|p_t(x,e) - p_t^0(x,e)| \leq ct^{-(D+\varepsilon)/2} \exp\left(-\frac{|x|_N^2}{ct}\right), \qquad x \in N,\ t \geq 1. \tag{1.14.4}$$

2. Lie groups of polynomial volume growth. Let L be a centered left invariant sub-Laplacian on a connected Lie group of polynomial volume growth G and let L_H be the associated homogenised operator.

Let also $p_t(x,y)$ and $p_t^H(x,y)$ be the heat kernels of L and L_H respectively.

Using the notation of section 1.4.7, we extend $p_t^H(x,y)$ to $S_1 \rtimes M_1 \cong G/C$ by setting $p_t^H((x,z),(y,w)) = p_t^H(x,y)$, $(x,z),(y,w) \in S_1 \rtimes M_1$ and then to G by setting $p_t^H(g,h) = p_t^H(\pi(g),\pi(h))$, $g,h \in G$.

If $X, Y, ..., Z$ are left invariant vector fields on $(S_1)_N$, then we also extend, in the same way, the kernel $XY...Zp_t^H(x,y)$ to G.

We have the following Berry-Esseen type of estimate (cf. [**Fe, Pe, JKO, Ko1, Ko2, Z**]):

THEOREM 1.14.5. *There are constants* $c > 0$ *such that*

$$|p_t(x,y) - p_t^H(x,y)| \leq ct^{-(D+1)/2}, \qquad x,y \in G,\ t \geq 1. \tag{1.14.5}$$

Combining the above result with (1.14.3) we have the following:

COROLLARY 1.14.6. *There are constants* $c, C_L \geq 0$ *such that*

$$|p_t(e,e) - C_L t^{-D/2}| \leq ct^{-(D+1)/2}, \qquad t \geq 1. \tag{1.14.6}$$

Also by interpolating (1.14.5) and (1.8.1) we have the following:

COROLLARY 1.14.7. *For all* $\varepsilon \in (0,1)$ *there is a constant* $c > 0$ *such that*

$$|p_t(x,y) - p_t^H(x,y)| \leq ct^{-(D+\varepsilon)/2} \exp\left(-\frac{|x^{-1}y|_G^2}{ct}\right), \qquad x,y \in G,\ t \geq 1. \tag{1.14.7}$$

Concerning the time derivative of $p_t(x,y)$ we have the following result:

THEOREM 1.14.8. *There is a constant* $c > 0$ *such that*

$$\left|\frac{\partial}{\partial t} p_t(x,y) - \frac{\partial}{\partial t} p_t^H(x,y)\right| \leq ct^{-(D+3)/2}, \qquad x,y \in G,\ t \geq 1. \tag{1.14.8}$$

3. A Berry-Esseen estimate for the spacial derivatives. For the space derivatives of $p_t(x,y)$, the situation is quite different. More precisely, if $X_1, ..., X_q$ is a convenient basis of $(S_1)_N$ and if $\psi^1, ..., \psi^{n_1}$ are the associated first order correctors, then we have the following result:

THEOREM 1.14.9. *For all $Y \in \mathfrak{g}$, there is a constant $c > 0$ such that*

$$|Yp_t(x,y) - Yp_t^H(x,y) - \sum_{1 \leq j \leq n_1} (Y\psi^j(x))\, X_j p_t^H(x,y)| \leq ct^{-(D+2)/2} \tag{1.14.9}$$

for all $x, y \in G$ and $t \geq 1$.

Combining (1.8.1), (1.13.2) and (1.14.9) we have the following:

COROLLARY 1.14.10. *For all $Y \in \mathfrak{g}$ and all $\varepsilon \in (0,1)$ there is a constant $c > 0$ such that*

$$\begin{aligned} |Yp_t(x,y) - Yp_t^H(x,y) - \sum_{1 \leq j \leq n_1} (Y\psi^j(x))\, X_j p_t^H(x,y)| \\ \leq ct^{-(D+1+\varepsilon)/2} \exp\left(-\frac{|x^{-1}y|_G^2}{ct}\right) \end{aligned} \tag{1.14.10}$$

for all $x, y \in G$ and $t \geq 1$.

The above results show that the role of the correctors is indeed essential. They also indicate how we can handle questions about the large time behavior of the spacial derivatives of the heat kernel.

1.15. Riesz transforms. Let $L = -\left(E_1^2 + ... + E_p^2\right) + E_0$ be a left invariant sub-Laplacian on a connected Lie group G of polynomial volume growth and let

$$L^{-n/2} = \frac{1}{\Gamma(n/2)} \int_0^\infty t^{(n/2)-1} e^{-tL} dt, \qquad n \in \mathbb{N}.$$

Let $\mathfrak{e}_1$ be the linear suspace of $\mathfrak{g}$ spanned by the vector fields $E_1, ..., E_p$ and by $\mathfrak{e}_2$ the linear sub-space of $\mathfrak{g}$ generated by the Lie brackets $[E_i, E_j]$, $1 \leq i, j \leq p$.

THEOREM 1.15.1. *Let us assume that L is centered and that $E_0 \in \mathfrak{e}_1 + \mathfrak{e}_2$. Then, the Riesz transform operators $E_i L^{-1/2}$ and $L^{-1/2} E_i$, $1 \leq i \leq p$ are bounded on L^r, for $1 < r < \infty$ and from L^1 to weak-L^1.*

The Riesz transforms are singular integral operators and so, to prove the above results, we apply the Calderon Zygmund theory (cf. [**S1, S3, CW**]). Their kernel is, in general, singular on the diagonal, as well as at infinity. The singularity near the diagonal can be treated by using results from the local theory of L. The geometry and the behavior of G and L at infinity is only used for the singularity at infinity.

In order to isolate these two singularities and treat them separately we consider the operators

$$L^{-n/2,0} = \frac{1}{\Gamma(n/2)} \int_0^1 t^{(n/2)-1} e^{-tL} dt \text{ and } L^{-n/2,\infty} = \frac{1}{\Gamma(n/2)} \int_1^\infty t^{(n/2)-1} e^{-tL} dt.$$

1. Local Riesz transforms. We have the following local result:

THEOREM 1.15.2. *Let us assume that $E_0 \in \mathfrak{e}_1 + \mathfrak{e}_2$. Then for all $Y_1, ..., Y_n \in \mathfrak{e}_1$, the Riesz transform operators*

$$R_{n,0} = Y_1 ... Y_n L^{-n/2,0} \quad \textit{and} \quad R_{n,0}^* = L^{-n/2,0} Y_1 ... Y_n$$

are bounded on L^p, for $1 < p < \infty$ and from L^1 to weak-L^1.

2. Riesz transforms at infinity. We have the following result:

THEOREM 1.15.3. *Let us assume that L is centered. Then, for all $Y \in \mathfrak{g}$, the Riesz transform operators*

$$R_\infty = YL^{-1/2,\infty}, \quad \textit{and} \quad R_\infty^* = L^{-1/2,\infty}Y$$

are bounded on L^p, for $1 < p < \infty$ and from L^1 to weak-L^1.

Note that in general the second order Riesz transforms $Y_1Y_2L^{-1,\infty}$ and $L^{-1,\infty}Y_1Y_2$, $Y_1, Y_2 \in \mathfrak{g}$ may be unbounded even on L^2 (cf. [**A2**]).

If G is nilpotent then there is no homogenisaton phenomena and so we can also consider higher order Riesz transforms:

THEOREM 1.15.4. *Let us assume that L is centered and that G is nilpotent. Then, for all $Y_1, ..., Y_n \in \mathfrak{g}$ the Riesz transform operators*

$$R_{n,\infty} = Y_1...Y_nL^{-n/2,\infty} \quad \textit{and} \quad R_{n,\infty}^* = L^{-n/2,\infty}Y_1...Y_n$$

are bounded on L^p, for $1 < p < \infty$ and from L^1 to weak-L^1.

Combining the theorems 1.15.2 and 1.15.4 we have the following:

COROLLARY 1.15.5. *Let L and E_0 be as in theorem 1.15.1 and let us assume that G is nilpotent. Then, for all $Y_1, ..., Y_n \in \mathfrak{e}_1$ the Riesz transform operators*

$$R_n = Y_1...Y_nL^{-n/2} \quad \textit{and} \quad R_n^* = L^{-n/2}Y_1...Y_n$$

are bounded on L^p, for $1 < p < \infty$ and from L^1 to weak-L^1.

2. The control distance and the local Harnack inequality

2.1. The control distance associated to a sub-Laplacian. To every left invariant sub-Laplacian $L = -\left(E_1^2 + ... + E_p^2\right) + E_0$ (or to every choice of left invariant vector fields $E_1, ..., E_p$ satisfying Hörmander's condition) on a connected Lie group G we can associate a so called control distance $d_L(.,.)$. This distance is left invariant, i.e $d_L(x,y) = d_L(e, x^{-1}y)$, $x, y \in G$ and it is defined as follows:
Let us denote by $\mathcal{C}$ the set of all absolutely continuous paths $\mathrm{c} : [0,1] \to G$, satisfying $\dot{\mathrm{c}}(t) = \sum_{1 \le i \le p} b_i(t)E_i\left(\mathrm{c}(t)\right)$, for almost all $t \in [0,1]$ and set

$$|\mathrm{c}| = \int_0^1 \left(b_1(t)^2 + ... + b_p(t)^2\right)^{1/2} dt.$$

We define

$$d_L(x,y) = \inf\left\{|\mathrm{c}| : \mathrm{c} \in \mathcal{C},\ \mathrm{c}(0) = x,\ \mathrm{c}(1) = y\right\}.$$

We denote by $B_r^L(x) = \{y \in N : d(x,y) < r\}$ the associated balls. We shall drop the index L when there is no risk of confusion.

The behavior of the balls $B_r(x)$ as $r \to \infty$ is actually a group invariant. More precisely, let us fix a compact neighborhood U of the identity element e of G and let $|.|_G$ be defined as in (1.1). Let us also consider a constant $c > 0$ such that

$$B_{1/c}(e) \subseteq U \subseteq B_c(e).$$

Then

$$B_{n/c}(e) \subseteq U^n \subseteq B_{cn}(e), \quad n \in \mathbb{N}. \tag{2.1.1}$$

This implies that there is $c > 0$ such that

$$\frac{1}{c}|x|_G \leq 1 + d(e,x) \leq c|x|_G, \qquad x \in G. \tag{2.1.2}$$

In contrast the behavior of the balls $B_r(x)$ as $r \to 0$ depends on the choice of the Hörmander vector fields $E_1, ..., E_p$. We can prove in particular (cf. [**NSW, VSC**]) that there is $d = d(E_1, ..., E_p) \in \mathbb{N}$ and a constant $c > 0$ such that for all $x \in G$,

$$\frac{1}{c}r^d \leq |B_r(x)| \leq cr^d, \quad 0 < r \leq 1. \tag{2.1.3}$$

2.2. A local Harnack inequality. Let $L = -\left(E_1^2 + ... + E_p^2\right) + E_0$ be a left invariant sub-Laplacian on a connected Lie group G. Then we have the following local Harnack inequality which is due to Bony [**Bo**] (see also [**VSC** ch. III]):

THEOREM 2.2.1. *Let V be a connected open subset of G, K a compact subset of V, $I = (a,b)$ and $a < t_1 \leq t_2 < s_1 \leq s_2 < b$. Then, for all $0 \leq k \in \mathbb{Z}$ and $0 \leq n \in \mathbb{Z}$ and all vector fields $Y_1, ..., Y_n \in \mathfrak{g}$ there is a constant $c_{k,n} > 0$ such for every positive solution of $\left(\frac{\partial}{\partial t} + L\right)u = 0$ in $I \times V$, we have*

$$\sup\left\{|\frac{\partial^k}{\partial t^k}Y_1, ..., Y_n u|; [t_1, t_2] \times K\right\} \leq c_{k,n} \inf\left\{u; [s_1, s_2] \times K\right\}. \tag{2.2.1}$$

This is a local result and the Lie group structure does not play any particular role in its proof.

Let us denote by $\mathfrak{e}_1$ the linear sub-space of $\mathfrak{g}$ generated by the vector fields $E_1, ..., E_p$ and by $\mathfrak{e}_2$ the linear sub-space of $\mathfrak{n}$ generated by the Lie brackets $[E_i, E_j]$, $1 \leq i,j \leq p$.

The following result is proved by using a local scaling and by observing that the rescaled vector fields satisfy Hörmander's condition in a uniform way (cf. [**V, VSC** ch. III]).

THEOREM 2.2.2. *Let L be as above and let us assume that $E_0 \in \mathfrak{e}_1 + \mathfrak{e}_2$. Then, for all $0 \leq k \in \mathbb{Z}$, $0 \leq n \in \mathbb{Z}$, for all $0 < \alpha < 1$, $0 < a < b < 1$ and all vector fields $Y_1, ..., Y_n \in \mathfrak{e}_1$ there is a constant $c_{k,n} > 0$ such that for all $t \in (0,1]$ and every positive solution u of $\left(\frac{\partial}{\partial t} + L\right)u = 0$ in $(0,t) \times B_{\sqrt{t}}(x)$, we have*

$$\sup\left\{|\frac{\partial^k}{\partial t^k}Y_1, ..., Y_n u(at,y)| : y \in B_{\sqrt{\alpha t}}\right\} \leq c_{k,n}\, t^{-k-n/2} \inf\left\{u(bt,y) : y \in B_{\sqrt{\alpha t}}\right\}. \tag{2.2.2}$$

3. The proof of the Harnack inequality from Varopoulos's theorem and propositions 1.6.3 and 1.6.4

In this section we shall give the proof of the following result, from Varopoulos's theorem 1.6.2 and by assuming propositions 1.6.3 and 1.6.4.

THEOREM 3.1. *Let L be a centered left invariant sub-Laplacian on a connected Lie group of polynomial volume growth G. Then for all $a, b > 0$ there is $\beta > \alpha > 1$, $c > 0$ and $\lambda > 0$ such that for all $r \geq 1$ and all $u \geq 0$ satisfying*

$$\left(\frac{\partial}{\partial t} + L\right) u = 0 \quad in \quad (0, (\beta + b^2)r^2) \times B_{cr}(e),$$

we have

$$\sup\left\{u; (\alpha r^2, (\alpha + a^2)r^2) \times B_{ar}(e)\right\} \leq \lambda \inf\left\{u; (\beta r^2, (\beta + b^2)r^2) \times B_{br}(e)\right\}. \tag{3.1}$$

In view of (2.2.1) and (2.2.2), theorem 1.6.1 follows from the above result.

The proof is inspired from Krylov and Safonov [**KS**]. In the first part of the proof we shall use theorem 1.6.2 and propositions 1.6.3 and 1.6.4, to prove the first growth lemma 3.1.1.

The second part of the proof consists of the proof of the second growth lemma 3.3.1. For the proof of this lemma we follow closely Krylov and Safonov [**KS**]. We use in particular certain covering lemmas.

A direct consequence of the second growth lemma is proposition 1.6.5 on the oscillation of the heat functions which, by a standard argument (see for example [**Law**]) implies (3.1).

The proof is long. This is, in part, due to the fact that we made an effort to write it in such a way that it can be easily adapted to convolution powers of densities on connected Lie groups and of probability measures on discrete groups of polynomial volume growth.

3.1. The first growth lemma.

If $B \subseteq \mathbb{R} \times G$, $A \subseteq B$ and $(t, x) \in B$ then, adopting the notation of [**KS**], we set

$$\Psi\left((t,x), A, B\right) = \\ \inf\left\{u(t,x) : u \geq 0, u(s,y) \geq 1 \text{ for } (s,y) \in A \text{ and } \left(\frac{\partial}{\partial t} + L\right) u = 0 \text{ in } B\right\}.$$

If $A' \subseteq B$ then we set

$$\Psi\left(A', A, B\right) = \inf\left\{\Psi\left((t,x), A, B\right) : (t,x) \in A'\right\}.$$

LEMMA 3.1.1 (first growth lemma). *For all $a > 1$ there is $c > a$ and $\delta, \xi \in (0,1)$ such that*

$$\Psi\left((a^{-2}r^2, a^2r^2) \times B_{ar}(e), A, (0, a^2r^2) \times B_{cr}(e)\right) > \delta \tag{3.1.1}$$

for all $r \geq 1$ and every measurable subset $A \subseteq (0, r^2) \times B_r(e)$, satisfying

$$|A| > \xi |(0, r^2) \times B_r(e)|.$$

As an immediate consequence of (3.1.1) and the local Harnack inequality 2.2.1, we have the following corollary:

COROLLARY 3.1.2. *For all $a > 1$ there is $c > a$, $\delta > 0$ and $m \in \mathbb{N}$ such that for all $r \geq 1$ and all $u \geq 0$ satisfying*

$$\left(\frac{\partial}{\partial t} + L\right) u = 0 \quad in \quad \big(0, (1+a^2)r^2\big) \times B_{cr}(e)$$

we have

$$\inf\left\{u; (1+a^{-2}r^2, 1+a^2r^2) \times B_{ar}(e)\right\} \geq \delta\ u(1,e)\ r^{-m}. \tag{3.1.2}$$

Moreover, if for some $1 \leq R \leq r$,

$$\inf\left\{u; (0, R^2) \times B_R(e)\right\} \geq 1$$

then

$$\inf\left\{u; (1+a^{-2}r^2, 1+a^2r^2) \times B_{ar}(e)\right\} \geq \delta\ \left(\frac{R}{r}\right)^{-m}. \tag{3.1.3}$$

3.2. Proof of the first growth lemma. The following lemma is an immediate consequence of the propositions (1.6.2) and (1.6.3).

LEMMA 3.2.1. *For all $a > 0$ there is $\delta > 0$ and $\xi \in (0,1)$ such that*

$$\int_A p_t(x,y)dy > \delta \tag{3.2.1}$$

for all $r \geq 1$, $(t,x) \in (a^{-2}r^2, a^2r^2) \times B_{ar}(e)$ and $A \subseteq B_r(e)$ satisfying

$$|A| > \xi |B_r(e)|.$$

Let us denote by $z(t)$ the diffusion process generated by the sub Laplacian L. Note that the transition function $P(t,x,A) = P[z(t+s) \in A | z(s) = x]$ of $z(t)$ satisfies

$$P(t,x,A) = \int_A p_t(x,y)dy.$$

Let us also denote by P_x, $x \in G$ the probabilities attached to the diffusion $z(t)$ and satisfying

$$P_x[z(0) = x] = 1 \quad \text{and} \quad P_x[z(t) \in A] = \int_A p_t(x,y)dy.$$

For every ball $B_r(x)$, we consider the stopping time

$$\tau_r^x = \inf\{t : z(t) \notin B_r(x)\}.$$

LEMMA 3.2.2. *For all $\varepsilon > 0$ there is a constant $c = c(\varepsilon) > 0$ such that*

$$P_x[\tau_{cr}^x \leq r^2] \leq \varepsilon, \qquad r \geq 1. \tag{3.2.2}$$

PROOF. By proposition 1.6.2, there is $\delta > 0$ and $c_\delta > 0$ such that

$$\int_{B_{c_\delta r}(x)} p_s(x,y)dy \geq \delta, \quad 0 < s \leq r^2$$

for all $r \geq 1$ and $x \in G$.

Let us fix $\varepsilon > 0$. Then, by proposition 1.6.3, there is $c > c_\delta$ such that

$$\int_{B_{cr}(x)^c} p_{r^2}(x,y)dy \leq \varepsilon\delta$$

for all $r \geq 1$ and $x \in G$.
We have

$$\begin{aligned}\varepsilon\delta \geq & \int_{B_{cr}(x)^c} p_{r^2}(x,y)dy \\ =& P_x[z(r^2) \in B_{cr}(x)^c] \\ \geq& E^{P_x}\left[P\left(r^2 - \tau^x_{2cr}, z(\tau^x_{2cr}), B_{cr}(x)^c\right) \mathbb{1}_{\{\tau^x_{2cr} \leq r^2\}}\right] \\ \geq& E^{P_x}\left[P\left(r^2 - \tau^x_{2cr}, z(\tau^x_{2cr}), B_{cr}(z(\tau^x_{2cr}))\right) \mathbb{1}_{\{\tau^x_{2cr} \leq r^2\}}\right] \\ =& E^{P_x}\left[P\left(r^2 - \tau^x_{2cr}, e, B_{cr}(e)\right) \mathbb{1}_{\{\tau^x_{2cr} \leq r^2\}}\right] \\ \geq& \delta P_x[\tau^x_{2cr} \leq r^2]\end{aligned}$$

and hence

$$P_x[\tau^x_{2cr} \leq r^2] \leq \varepsilon$$

which proves the lemma.

PROOF OF LEMMA 3.1.1. Let $t_0 \in (0, \frac{1}{2}a^{-2}r^2)$ and let us denote by A_{t_0} the section

$$A_{t_0} = A \cap \{t_0\} \times B_r(e).$$

Let $c > 1 + a$ and let τ^x_{cr} be as in the lemma 3.2.2.

Let also $u \geq 0$ satisfying $u(s,y) \geq 1$, for $(s,y) \in A$ and $\left(\frac{\partial}{\partial t} + L\right)u = 0$ in $(0, a^2r^2) \times B_{2cr}(e)$. Then

$$\begin{aligned}u(t,x) \geq& E^{P_x}[u(t_0, z(t-t_0)); \tau^x_{cr} > t - t_0] \\ \geq& E^{P_x}[\mathbb{1}_{A_{t_0}}(z((t-t_0)); \tau^x_{cr} > t - t_0] \\ =& \int_{A_{t_0}} p_{t-t_0}(x,y)dy - P_x[\tau^x_{cr} \leq t - t_0]\end{aligned} \tag{3.2.3}$$

for all $(t,x) \in (a^{-2}r^2, a^2r^2) \times B_{ar}(e)$.

Now, by lemma 3.2.1, there is $\delta > 0$ and $\xi_0 \in (0,1)$ such that

$$\int_{A_{t_0}} p_{t-t_0}(x,y)dy > 2\delta$$

for all $(t,x) \in (a^{-2}r^2, a^2r^2) \times B_{ar}(e)$, if A_{t_0} satisfies

$$|A_{t_0}| > \xi_0 |B_r(e)|. \tag{3.2.4}$$

If we assume that $|A| > \xi|(0,r^2) \times B_r(e)|$, with $\xi \in [\xi_0, 1)$ near enough to 1, then A will always have a section A_{t_0} with $t_0 \in (0, \frac{1}{2}a^{-2}r^2)$ and satisfying (3.2.4).

Also by lemma 3.2.2 there is $c_\delta > 0$ such that

$$P_x[\tau^x_{c_\delta r} \leq t - t_0] < \delta$$

for all $t \in (a^{-2}r^2, a^2r^2)$.

So, if we take $c > 1 + a + c_\delta$ then by (3.2.3)

$$\begin{aligned} u(t,x) \geq & \int_{A_{t_0}} p_{t-t_0}(x,y)dy - P_x[\tau^x_{cr} \leq t - t_0] \\ \geq & 2\delta - \delta = \delta \end{aligned}$$

for all $(t,x) \in (a^{-2}r^2, a^2r^2) \times B_{ar}(e)$ and the lemma follows.

3.3. The second growth lemma.

LEMMA 3.3.1 (second growth lemma). *For all $b > 0$, there is $\beta > 1$, $c > 1$, $\delta > 0$ and $m \in \mathbb{N}$ such that such that*

$$\Psi\Big((\beta r^2, (\beta + b^2)r^2) \times B_{br}(e), A, \big(0, (\beta + b^2)r^2\big) \times B_{cr}(e)\Big) > \delta \left(\frac{|A|}{|(1, 1 + r^2) \times B_r(e)|} \right)^m \tag{3.3.1}$$

for all $r \geq 1$ and every measurable subset $A \subseteq (1, 1 + r^2) \times B_r(e)$.

The above lemma will actually be a consequence of the following:

LEMMA 3.3.2. *For all $b > 0$, there is $\beta > 1$, $c > 1$, $\delta > 0$, $\delta' > 0$, $\theta > 0$ and $m \in \mathbb{N}$ such that such that for all $r \geq 1$ and every measurable subset $A \subseteq (1, r^2) \times B_r(e)$ either*

$$\Psi\Big((\beta r^2, (\beta + b^2)r^2) \times B_{br}(e), A, \big(0, (\beta + ab2)r^2\big) \times B_{cr}(e)\Big) > \delta \left(\frac{|A|}{|(1, 1 + r^2) \times B_r(e)|} \right)^m, \tag{3.3.2}$$

or there is a measurable subset $A_0 \subseteq (1, 1 + r^2) \times B_r(e)$ such that

$$|A_0| > (1 + \theta)|A| \tag{3.3.3}$$

and

$$\Psi\big(A_0, A, \big(0, 1 + r^2\big) \times B_{cr}(e)\big) > \delta'. \tag{3.3.4}$$

3.4. Proof of the lemma 3.3.2. We shall use the notation

$$Q(s,t,x) = (t - \frac{1}{2}s^2, t + \frac{1}{2}s^2) \times B_s(x).$$

Let us fix $r_0 > 3$ and $k > r_0^2$.

If $r^2 \leq 6k$, then (3.3.2) follows from (3.1.2) (or the local Harnack inequality (2.2.1)).

So we shall assume from now on that

$$r^2 > 6k.$$

We set

$$A_1 = A \cap (3k, 1 + r^2 - 3k) \times B_r(e) \quad \text{and} \quad A_2 = A \setminus A_1.$$

Let $\alpha \in (0,1)$ to be determined later.

Case I : $|A_1| \leq \alpha |A|$.

Then $|A_2| > (1-\alpha)|A|$ and hence

$$\begin{aligned}\frac{|A|}{|(1,r^2) \times B_r(e)|} &\leq \frac{1}{1-\alpha} \frac{|A_2|}{|(1,1+r^2) \times B_r(e)|} \\ &\leq \frac{1}{1-\alpha} \frac{|(0,3k] \times B_r(e)| + |[1+r^2-3k, 1+r^2) \times B_r(e)|}{|(1,1+r^2) \times B_r(e)|} \\ &= \frac{6k}{1-\alpha} \frac{1}{r^2}.\end{aligned}$$

If $A \neq \emptyset$ then there is $(t,x) \in (1, 1+r^2) \times B_r(e)$ such that $u(t,x) \geq 1$ and so (3.3.2) follows from (3.1.2).

Case II : $|A_1| > \alpha |A|$.

By the local Harnack inequality (2.2.1), there is $c > r_0$ and $\delta_1 > 0$ such that

$$\Psi\Big(Q(r_0, 2k, x), \{(k,x)\}, (0, 2k + \frac{1}{2} r_0^2) \times B_{cr_0}(x)\Big) > \delta_1. \tag{3.4.1}$$

Let

$$A_k = \{(t,x) : (t-k, x) \in A_1\}.$$

Then

$$A_k \subseteq (4k, 1 + r^2 - 2k) \times B_r(e) \quad \text{and} \quad |A_k| = |A_1| > \alpha |A|.$$

We set

$$A_{\delta_1} = \cup_{(t,x) \in A_k} Q(r_0, t, x) \cap (1, r^2) \times B_r(e).$$

Then of course $A_k \subseteq A_{\delta_1}$ and by (3.4.1)

$$\Psi\big(A_{\delta_1}, A, (0, 1+r^2) \times B_{cr}(e)\big) > \delta_1. \tag{3.4.2}$$

Let us fix $\eta > 1$.

Then, by lemma 3.4.1, there is $\delta_2, \xi \in (0,1)$ and $c > \eta$ such that

$$\Psi\left((\eta^{-2}s^2, \eta^2 s^2) \times B_{\eta s}(e), V, (0, \eta^2 s^2) \times B_{cs}(e)\right) > \delta_2 \tag{3.4.3}$$

for all $s \geq 1$ and every measurable subset $V \subseteq (0, s^2) \times B_s(x)$, satisfying

$$|V| \geq \xi |(0, s^2) \times B_s(e)|.$$

We consider the set of balls

$$\mathcal{Q} = \big\{Q(s,t,x) \subseteq (1,r^2) \times B_r(e) : s \geq 1, s + d(e,x) \leq r \text{ and } |Q \cap A_{\delta_1}| \geq \xi |Q|\big\}$$

We have put the technical condition $s + d(e,x) \leq r$ in order not to make use of any further properties of the control distance $d(.,.)$.

We set

$$W = \cup_{Q \in \mathcal{Q}} Q.$$

Then of course $A_k \subseteq W$.

LEMMA 3.4.1. *Either*

$$(1, 1 + r^2) \times B_r(e) \in \mathcal{Q},$$

or there is $\theta_1 = \theta_1(\xi) > 0$ *(i.e.* θ_1 *depends only on* ξ *and the group* G*) such that*

$$|W| > (1 + \theta_1)|A_k| \tag{3.4.4}$$

and hence

$$|W| > (1 + \theta_1)\alpha|A|. \tag{3.4.5}$$

The proof of the above lemma will be given later.

If $(1, 1 + r^2) \times B_r(e) \in \mathcal{Q}$ then (3.3.2) follows from lemma 3.1.1 and corollary 3.1.2. So we shall assume that

$$(1, 1 + r^2) \times B_r(e) \notin \mathcal{Q}.$$

By using (3.4.3) repeatedly (or by using corollary 3.1.2), we can see that there is $\delta_2 > 0$ and $m \in \mathbb{N}$ such that for all $\sigma \geq 1$,

(3.4.6)

$$\Psi\Big((s^2, \sigma^2\eta^2 s^2) \times B_{\eta s}(e), (\eta^{-2}s^2, \eta^2 s^2) \times B_{\eta s}(e), (0, \sigma^2\eta^2 s^2) \times B_{cs}(e)\Big) \geq \delta_2 \sigma^{-m}.$$

Let us fix $\sigma \geq 1$ such that

$$(1 + \theta_1)\frac{\sigma^2\eta^2 - 1}{\sigma^2\eta^2} > 1 + \frac{\theta_1}{2}. \tag{3.4.7}$$

If $Q = (t, t + s^2) \times B_s(x) \in \mathcal{Q}$ then we set

$$Q^1 = (t, t + \sigma^2\eta^2 s^2) \times B_s(x), \quad \mathcal{Q}^1 = \{Q^1, Q \in \mathcal{Q}\}$$

and

$$W^1 = \cup_{Q^1 \in \mathcal{Q}^1} Q^1.$$

If $Q^1 = (t, t + \sigma^2\eta^2 s^2) \times B_s(x) \in \mathcal{Q}^1$ then we set

$$Q^2 = (t + s^2, t + \sigma^2\eta^2 s^2) \times B_s(x), \quad \mathcal{Q}^2 = \{Q^1, Q \in \mathcal{Q}^1\}$$

and

$$W^2 = \cup_{Q^2 \in \mathcal{Q}^2} Q^2.$$

It follows from (3.4.2), (3.4.3) and (3.4.6) that if the constant c is chosen large enough then there is $\delta_3 > 0$ such that

$$\Psi\big(W^2, A, (0, c^2 r^2) \times B_{cr}(e)\big) > \delta_3. \tag{3.4.8}$$

Let

$$\gamma = \frac{|A|}{|(1, 1 + r^2) \times B_r(e)|}$$

and let $\omega \in (0, 1)$ to be determined later.

Case IIa : $|W^2 \setminus (1, 1 + r^2) \times B_r(e)| \geq \omega|A|$.

This assumption implies that

$$|W^2 \setminus (1, 1 + r^2) \times B_r(e)| \geq \omega\gamma|(1, 1 + r^2) \times B_r(e)|.$$

So there is a ball

$$Q^2 = \left(t+s^2, t+\sigma^2\eta^2 s^2\right) \times B_s(x) \in \mathcal{Q}^2$$

such that

$$\sigma^2\eta^2 s^2 \geq \omega\gamma r^2.$$

Now, by corollary 3.1.2, for all $a_1 > 0$, there is $c \geq a_1$, $\delta_4 > 0$ and $m \in \mathbb{N}$ such that for all $R \geq \sigma\eta s$,

$$\begin{aligned}
&\Psi\left((t+R^2, t+(1+a_1^2)R^2) \times B_{a_1 R}(x), Q^2, (t, t+(1+a_1^2)R^2) \times B_{cR}(x)\right) \\
&\geq \delta_4 \left(\frac{\sigma\eta s}{R}\right)^m \\
&\geq \delta_4 \omega^{m/2} \left(\frac{r}{R}\right)^m \gamma^{m/2}.
\end{aligned} \tag{3.4.9}$$

The lemma follows from (3.4.9) above, by taking a_1 large enough and by replacing R by an appropriate multiple of r.

Case IIb : $|W^2 \setminus (1, 1+r^2) \times B_r(e)| < \omega|A|$.

Let us first observe that since $W \subseteq W^1$ we have

$$|W| \leq |W^1|. \tag{3.4.10}$$

The following lemma is the analog of the lemma 2.3 of [**KS** p.158].

LEMMA 3.4.2.

$$|W^1| \leq \frac{\sigma^2\eta^2}{\sigma^2\eta^2 - 1}|W^2|. \tag{3.4.11}$$

The proof of the above lemma will be given later.
Combining (3.4.11) with (3.4.10) and (3.4.7) we have

$$\begin{aligned}
|W^2| \geq& \frac{\sigma^2\eta^2 - 1}{\sigma^2\eta^2}|W| \geq \frac{\sigma^2\eta^2 - 1}{\sigma^2\eta^2}(1+\theta_1)\alpha|A| \\
\geq& (1+\frac{\theta_1}{2})\alpha|A|.
\end{aligned}$$

We set

$$A_0 = W^2 \cap (1, 1+r^2) \times B_r(e).$$

Then we have

$$\begin{aligned}
|A_0| =& |W^2 \cap (1, 1+r^2) \times B_r(e)| \\
=& |W^2| - |W^2 \setminus (1, 1+r^2) \times B_r(e)| \\
\geq& |W^2| - \omega|A| \geq (1+\frac{\theta_1}{2})\alpha|A| - \omega|A| \\
\geq& \left[(1+\frac{\theta_1}{2})\alpha - \omega\right]|A|.
\end{aligned}$$

It follows that if we chose $\alpha \in (0,1)$ so that

$$(1+\frac{\theta_1}{2})\alpha > 1 + \frac{\theta_1}{4}$$

and

$$\omega \in (0, \frac{\theta_1}{8}],$$

then we shall have

$$|A_0| > (1 + \frac{\theta_1}{8})|A|$$

which proves (3.3.3).

3.5. Proof of lemma 3.4.1. Let us assume that $(1, 1 + r^2) \notin \mathcal{Q}$ and let us fix an $\varepsilon > 0$ such that

$$\xi < (1 + \varepsilon)\xi < 1. \tag{3.5.1}$$

We set

$$\mathcal{Q}^0 = \{Q \in \mathcal{Q} : \xi|Q| \le |Q \cap A_{\delta_1}| \le (1 + \varepsilon)|Q|\}.$$

We have the following :

LEMMA 3.5.1.

$$\cup_{Q^0 \in \mathcal{Q}^0} Q_0 = \cup_{Q \in \mathcal{Q}} Q = W. \tag{3.5.2}$$

PROOF. We shall use the notation

$$D_s(t, x) = (t - s^2, t) \times B_s(x).$$

We shall prove that for all $D \in \mathcal{Q}$ and all $(\tau, y) \in D$ there is $D^0 \in \mathcal{Q}^0$ such that $(\tau, y) \in D^0$. This of course will imply the lemma.

So let us fix a ball $D_s(t, x) = (t - s^2, t) \times B_s(x) \in \mathcal{Q}$ and a point $(\tau, y) \in D_s(t, x)$. If

$$\xi|D_s(t, x)| \le |D_s(t, x) \cap A_{\delta_1}| \le (1 + \varepsilon)\xi|D_s(t, x)|,$$

then there is nothing to prove. So let us assume that

$$|D_s(t, x) \cap A_{\delta_1}| > (1 + \varepsilon)\xi|D_s(t, x)|. \tag{3.5.3}$$

Since $d(x, y) < s$, by chosing a slightly smaller value for s if necessary, we can assume that $s + d(e, x) < r$.

If there is $\sigma \in [s + d(e, x), r)$ such that

$$\xi|(1, 1 + \sigma^2) \times B_\sigma(e)| \le |(1, 1 + \sigma^2) \times B_\sigma(e) \cap A_{\delta_1}| \le (1 + \varepsilon)\xi|(1, 1 + \sigma^2) \times B_\sigma(e)|$$

then we just take $D^0 = (1, 1 + \sigma^2) \times B_\sigma(e)$.

So, let us assume that this is not the case, i.e. that

$$|(1, 1 + \sigma^2) \times B_\sigma(e) \cap A_{\delta_1}| < \xi|(1, 1 + \sigma^2) \times B_\sigma(e)| \tag{3.5.4}$$

for all $\sigma \in [s + d(e, x), r)$.

We set

$$\zeta = r - d(e, x) - s \quad \text{and} \quad T = d(e, x) + \frac{\zeta}{4}$$

and we consider an admissible curve (cf. section 2.1) $\mathrm{c} : [0, T] \to N$ such that $\mathrm{c}(0) = e$ and $\mathrm{c}(T) = x$.

For every $\sigma \in [0, T]$, we set $v(\sigma) = s + T - \sigma$.

If $v(0)^2 \le t - 1$, then we set

$$D^\sigma = \left(t - v(\sigma)^2, t\right) \times B_{v(\sigma)}\left(\mathrm{c}(\sigma)\right), \qquad \sigma \in [0, T].$$

If $v(0)^2 > t-1$, then we set

$$D^\sigma = \begin{cases} \left(t - v(\sigma)^2, t\right) \times B_{v(\sigma)}\left(c(\sigma)\right), & \text{for } v(\sigma)^2 \leq t-1 \\ \left(1, 1 + v(\sigma)^2\right) \times B_{v(\sigma)}\left(c(\sigma)\right), & \text{for } v(\sigma)^2 > t-1. \end{cases}$$

We observe that

$$\begin{aligned} d(c(\sigma), y) \leq & d(c(\sigma), x) + d(x, y) < d(c(\sigma), x) + s \\ \leq & T - \sigma + s = v(\sigma) \end{aligned}$$

and hence $(\tau, y) \in D^\sigma$ for all $\sigma \in [0, T]$.

Also by (3.5.3) and (3.5.4) there is $\sigma_0 \in [0, T]$ such that

$$\xi |D^{\sigma_0}| \leq |D^{\sigma_0} \cap A_{\delta_1}| \leq (1 + \varepsilon)\xi |D^{\sigma_0}|$$

and the lemma follows.

If $Q = Q(s,t,x) = (t - \frac{1}{2}s^2, t + \frac{1}{2}s^2) \times B_s(x)$, then we shall denote

$$Q^* = Q^*(s,t,x) = (t - \frac{25}{2}s^2, t + \frac{25}{2}s^2) \times B_{5s}(x).$$

LEMMA 3.5.2. *There is a finite sequence of balls $Q^1, Q^2, Q^3, ..., Q^n$ such that*
1. $Q^i \in \mathcal{Q}^0$, $1 \leq i \leq n$,
2. $Q^i \cap Q^j = \emptyset$, $i \neq j$, $1 \leq i, j \leq n$ *and*
3. $W \subseteq \cup_{i=1}^n Q^{i*}$.

PROOF. Let $\mathcal{A}^0 = \mathcal{Q}^0$ and set

$$s_0 = \sup\{\text{radius}(Q) : Q \in \mathcal{A}^0\}$$

where radius $(Q(s,t,x)) = s$.
We chose as Q^1 any ball in $\mathcal{Q}^0$ satisfying

$$\tau_1 = \text{radius}(Q^1) \geq \frac{s_0}{2}. \tag{3.5.5}$$

Assume that we have chosen the balls $Q^1, Q^2, ..., Q^k$. Then in order to chose the ball Q^{k+1} let us set

$$\mathcal{A}^k = \{Q \in \mathcal{Q}^0 : Q \cap Q^i = \emptyset, \quad i = 1, 2, ..., k\}.$$

If $\mathcal{A}^k = \emptyset$ then we stop there. If not, then we set

$$s_k = \sup\{\text{radius}(Q) : Q \in \mathcal{A}^k\}$$

and we chose as Q^{k+1} any ball in $\mathcal{A}^k$ satisfying

$$\tau_{k+1} = \text{radius}(Q^{k+1}) \geq \frac{s_k}{2}. \tag{3.5.6}$$

Note that since the chosen balls are disjoint and of radius ≥ 1, we will eventually have $\mathcal{A}^n = \emptyset$ and this proccess will stop. We obtain in this way a finite sequence $Q^1, Q^2, ..., Q^n$ of balls satisfying (1) and (2).

It remains to prove that (3) is also satisfied, i.e. that $W \subseteq \cup_{i=1}^n Q^{i*}$. For this, it is enough to prove that if Q is any ball $Q \in \mathcal{Q}^0$ then there is $i_0 \in \{1, 2, ..., n\}$ such that

$$Q \subseteq Q^{i_0 *}. \tag{3.5.7}$$

Since $\mathcal{A}^n = \emptyset$ and $Q \in \mathcal{A}^0 = \mathcal{Q}^0$, there is $i_0 \in \{1, 2, ..., n\}$ such that $Q \in \mathcal{A}^{i_0 - 1}$ and $Q \notin \mathcal{A}^{i_0}$. Let s =radius(Q). Then, by (3.5.5) and (3.5.6) we have $Q \cap Q^{i_0} \neq \emptyset$ and $s \leq 2\tau_{i_0}$ and hence $Q \in Q^{i_0 *}$. This proves (3.5.7) and the lemma follows.

PROOF OF LEMMA 3.4.1. Let us first observe that by (1.2) there is a constant $c \geq 1$ such that

$$\frac{1}{c} \leq \frac{|Q^*(s,t,x)|}{|Q(s,t,x)|} \leq c \tag{3.5.8}$$

for all $s \geq 1$.
We have

$$\begin{aligned}\frac{|W|}{|A_k|} &= \frac{|A_k| + |W \setminus A_k|}{|A_k|} = 1 + \frac{|W \setminus A_{\delta_1}|}{|A_k|} \\ &\geq 1 + \frac{|W \setminus A_{\delta_1}|}{|W|} \geq 1 + \frac{|W \setminus A_{\delta_1}|}{|\cup_{i=1}^n Q^{i*}|} \\ &\geq 1 + \frac{|W \setminus A_{\delta_1}|}{\sum_{i=1}^n |Q^{i*}|} \geq 1 + \frac{|W \setminus A_{\delta_1}|}{c\sum_{i=1}^n |Q^i|} \\ &\geq 1 + \frac{|\cup_{i=1}^n Q^i \setminus A_{\delta_1}|}{c\sum_{i=1}^n |Q^i|} = 1 + \frac{\sum_{i=1}^n |Q^i \setminus A_{\delta_1}|}{c\sum_{i=1}^n |Q^i|}.\end{aligned} \tag{3.5.9}$$

Since

$$\xi|Q^i| \leq |Q^i \cap A_{\delta_1}| \leq (1+\varepsilon)\xi|Q^i|,$$

we have

$$|Q^i \setminus A_{\delta_1}| = |Q^i| - |Q^i \cap A_{\delta_1}| \geq |Q^i| - (1+\varepsilon)\xi|Q^i| = [1-(1+\varepsilon)\xi]\,|Q^i|. \tag{3.5.10}$$

Combining (3.5.9) and (3.5.10) we have that

$$\frac{|W|}{|A_k|} \geq 1 + \frac{\sum_{i=1}^n [1-(1+\varepsilon)\xi]\,|Q^i|}{c\sum_{i=1}^n |Q^i|} = 1 + \frac{1-(1+\varepsilon)\xi}{c}$$

which proves the lemma.

3.6. Proof of lemma 3.4.2. We set

$$W_x^1 = W^1 \cap \mathbb{R} \times \{x\} \quad \text{and} \quad W_x^2 = W^2 \cap \mathbb{R} \times \{x\}$$

for $x \in B_r(e)$.
It is enough to prove that

$$|W_x^1| \leq \frac{\sigma^2\eta^2}{\sigma^2\eta^2 - 1}|W_x^2|. \tag{3.6.1}$$

We shall need the following lemma from [**KS**]:

LEMMA 3.6.1 (cf. [**KS** lemma 2.2 on p. 157]). *Let $\kappa \geq 1$, let*

$$\mathcal{A} = \{(t_1, t_2) \subseteq \mathbb{R} : -\infty \leq t_1 < t_2 \leq \infty\}$$

and let g a function $g : \mathcal{A} \to \mathcal{A}$ satisfying

1. $|g(I)| \leq \kappa|I|, \quad I \in \mathcal{A}$ *and*
2. $g(I_1) \subseteq g(I_2)$, *if* $I_1 \subseteq I_2$.

Then, for all $\mathcal{B} \subseteq \mathcal{A}$,

$$|\cup_{I\in\mathcal{B}}\, g(I)| \le \kappa |\cup_{I\in\mathcal{B}}\, I|.$$

PROOF. The proof is taken from [**KS**] and is given for reasons of completeness.

The set $\cup_{I\in\mathcal{B}} I$ is open and hence it can be represented as the union $\cup_{I\in\mathcal{B}} I = \cup_n I_n$ of certain nonintersecting intervals I_n. So we have

$$|\cup_{I\in\mathcal{B}}\, g(I)| \le |\cup_n \cup_{I\subseteq I_n} g(I)| \le |\cup_n g(I_n)| \le \sum_n |g(I_n)|$$
$$\le \sum_n \kappa |I_n| = \kappa \sum_n |\cup_{I\in\mathcal{B}}\, I|$$

which proves the lemma.

Now (3.6.1) follows from the above lemma, by taking

$$\kappa = \frac{\sigma^2\eta^2}{\sigma^2\eta^2 - 1}, \qquad \mathcal{B} = \{\mathbb{R} \times \{x\} \cap Q^2 : Q^2 \in \mathcal{Q}^2\}$$

and by setting

$$g\left((t_1, t_2)\right) = \left(t_2 - \kappa\left(t_2 - t_1\right), t_2\right).$$

3.7. The Oscillation of the heat functions. Proposition 1.6.5 is a consequence of the following:

PROPOSITION 3.7.1. *Let* b, β *and* c *be as in lemma 3.3.1. Then there is* $\alpha \in (0,1)$ *such that*

$$\text{(3.7.1)} \quad \operatorname{Osc}\left(u, (\beta r^2, (\beta + b^2)r^2) \times B_{br}(e)\right) \le \alpha \operatorname{Osc}\left(u, (0, (\beta + b^2)r^2) \times B_{cr}(e)\right)$$

for all $r \ge 1$ *and all functions* u *satisfying*

$$\left(\frac{\partial}{\partial t} + L\right) u = 0 \quad \text{in} \quad (0, (\beta + b^2)r^2) \times B_{cr}(e).$$

PROOF. Let us fix a function u satisfying

$$\left(\frac{\partial}{\partial t} + L\right) u = 0 \quad \text{in} \quad (0, (\beta + b^2)r^2) \times B_{cr}(e).$$

and let

$$m_1 = \inf\left\{u; (0, (\beta + b^2)r^2) \times B_{cr}(e)\right\}, \quad M_1 = \sup\left\{u; (0, (\beta + b^2)r^2) \times B_{cr}(e)\right\}$$

$$m_2 = \inf\left\{u; (\beta r^2, (\beta + b^2)r^2) \times B_{br}(e)\right\},$$
$$M_2 = \sup\left\{u; (\beta r^2, (\beta + b^2)r^2) \times B_{br}(e)\right\}.$$

Let also

$$A = \left\{(t,x) \in (1, r^2) \times B_r(e) : u(t,x) \ge \frac{1}{2}(m_1 + M_1)\right\}.$$

Case I : $|A| \ge \frac{1}{2}|(1, 1 + r^2) \times B_r(e)|$.

Let us observe that

$$\frac{1}{2}(m_1 + M_1) - m_1 = \frac{1}{2}(M_1 - m_1).$$

So the function

$$v = (u - m_1)\left(\frac{1}{2}(M_1 - m_1)\right)^{-1} = 2(u - m_1)(M_1 - m_1)^{-1}$$

satisfies $v(t, x) \geq 1$ for $(t, x) \in A$.
So, by (3.3.1) there is $\delta > 0$ such that

$$\inf\{v; (\beta r^2, (\beta + b^2)r^2) \times B_{br}(e)\} > \delta.$$

This implies that

$$m_2 > m_1 + \frac{1}{2}\delta(M_1 - m_1).$$

and hence

$$\begin{aligned} M_2 - m_2 \leq & M_2 - m_1 - \frac{1}{2}\delta(M_1 - m_1) \\ \leq & M_1 - m_1 - \frac{1}{2}\delta(M_1 - m_1) = (1 - \frac{1}{2}\delta)(M_1 - m_1). \end{aligned}$$

which proves (3.7.1).

Case II : $|A| < \frac{1}{2}|(1, 1 + r^2) \times B_r(e)|$.

Then we consider the function $v = M_1 - u$ and we are in the previous case.

3.8. End of the proof of theorem 3.1. We shall use again the notation

$$D_r(t, x) = (t - r^2, t) \times B_r(x).$$

Let us fix $b > 0$. Then by corollary 3.1.2 there is $\beta > 1$, $c > 1$, $\delta > 0$ and $m \in \mathbb{N}$ such that such that

1. for all $r \geq 1$ and all $(t, x) \in (1, r^2) \times B_r(e)$,

(3.8.1) $$\Psi\left(\left(\beta r^2, (\beta + b^2)r^2\right) \times B_{br}(e), (t, x), \left(0, (\beta + b^2)r^2\right) \times B_{cr}(e)\right) > \delta r^{-m}.$$

2. for all $r \geq 1$ and every ball $D_{\varepsilon r} \subseteq (1, r^2) \times B_r(e)$,

(3.8.2) $$\Psi\left(\left(\beta r^2, (\beta + b^2)r^2\right) \times B_{br}(e), D_{\varepsilon r}(t, x), \left(0, (\beta + b^2)r^2\right) \times B_{cr}(e)\right) > \delta \varepsilon^m.$$

We shall prove that there is a constant $\lambda > 0$ such that

$$\sup\left\{u; (\frac{3}{4}r^2, r^2) \times B_{r/2}(e)\right\} \leq \lambda$$

for all $r \geq 4$ all functions $u \geq 0$ satisfying

$$\left(\frac{\partial}{\partial t} + L\right) u = 0 \quad \text{in} \quad (0, (\beta + b^2)r^2) \times B_{cr}(e)$$

$$\inf\left\{u; (\beta r^2, (\beta + b^2)r^2) \times B_{br}(e)\right\} = 1.$$

This of course will prove the theorem.

Let us observe that by (3.7.1)

(3.8.3) $$\sup\left\{u; (1, r^2) \times B_r(e)\right\} \leq \frac{1}{\delta}r^m$$

and that by (3.8.2)

$$\inf\{u; D_{\varepsilon r}(t,x)\} \leq \frac{1}{\delta}\varepsilon^{-m} \tag{3.8.4}$$

for every ball $D_{\varepsilon r}(t,x) \subseteq (1,r^2) \times B_r(e)$.
Furthermore, by (3.7.2), there is $C > 1$ and $\alpha \in (0,1)$ such that

$$\operatorname{Osc}\big(u, D_s(t,x)\big) \leq \alpha \operatorname{Osc}\big(u, D_{Cs}(t,x)\big) \tag{3.8.5}$$

for all $D_{Cs}(t,x) \subseteq (0,r^2) \times B_r(e)$.

Let us fix $\zeta \in (\alpha, 1)$ satisfying

$$\zeta^m > \alpha > \zeta^{m+1}. \tag{3.8.6}$$

Let also fix $k \in \mathbb{N}$ such that

$$\zeta^k \frac{1}{1-\zeta} < \frac{1}{4}. \tag{3.8.7}$$

We set

$$\lambda = \frac{1}{\delta}\alpha^{\log C/\log\zeta} + \frac{1}{\delta}C^m\zeta^{-km}\frac{1}{1-\alpha\zeta^{-m}}.$$

We shall prove that

$$\sup\left\{u; (\frac{3}{4}r^2, r^2) \times B_{r/2}(e)\right\} \leq \lambda. \tag{3.8.8}$$

Case I : $\frac{1}{C}\zeta^k r \leq 1$.

Then $r \leq C\zeta^{-k}$ and hence, by (3.8.3),

$$\sup\{u; (1,r^2) \times B_r(e)\} \leq \frac{1}{\delta}C^m\zeta^{-km}$$

which proves (3.8.8).

Case II : $\frac{1}{C}\zeta^k r > 1$.

Let us assume the contrary, i.e. that

$$\sup\left\{u; (\frac{3}{4}r^2, r^2) \times B_{r/2}(e)\right\} > \lambda.$$

We shall reach a contradiction.

Let $(t_0, x_0) \in (\frac{3}{4}r^2, r^2) \times B_{r/2}(e)$ such that

$$u(t_0,x_0) > \lambda.$$

We consider a ball $D_{\frac{1}{C}\zeta^k r}(s_0,y_0) \subseteq (\frac{3}{4}r^2, r^2) \times B_{r/2}(e)$ such that

$$(t_0,x_0) \in D_{\frac{1}{C}\zeta^k r}(s_0,y_0).$$

Then by (3.8.4)

$$\inf\{u; D_{\frac{1}{C}\zeta^k r}(s_0,y_0) \leq \frac{1}{\delta}C^m\zeta^{-km}$$

and hence

$$\operatorname{Osc}\big(u, D_{\frac{1}{C}\zeta^k r}(s_0,y_0)\big) \leq \lambda - \frac{1}{\delta}C^m\zeta^{-km}.$$

By (3.8.5) we have

$$\begin{aligned}\operatorname{Osc}\left(u, D_{\zeta^k r}(s_0, y_0)\right) \geq & \frac{1}{\alpha} \operatorname{Osc}\left(u, D_{\frac{1}{C}\zeta^k r}(s_0, y_0)\right) \\ \geq & \frac{\lambda}{\alpha} - \frac{1}{\alpha}\frac{1}{\delta} C^m \zeta^{-km}\end{aligned}$$

and hence there is $(t_1, x_1) \in D_{\zeta^k r}(s_0, y_0)$ such that

$$u(t_1, x_1) \geq \frac{\lambda}{\alpha} - \frac{1}{\alpha}\frac{1}{\delta} C^m \zeta^{-km}.$$

If $\frac{1}{C}\zeta^{k+1} r \leq 1$ then we stop there. If not, then we consider a ball

$$D_{\frac{1}{C}\zeta^{k+1} r}(s_1, y_1) \subseteq D_{\zeta^k r}(s_0, y_0)$$

such that

$$(t_1, x_1) \in D_{\frac{1}{C}\zeta^{k+1} r}(s_1, y_1).$$

Again by (3.8.4)

$$\inf\{u; D_{\frac{1}{C}\zeta^{k+1} r}(s_1, y_1)\} \leq \frac{1}{\delta} C^m \zeta^{-(k+1)m}$$

and hence

$$\operatorname{Osc}\left(u, D_{\frac{1}{C}\zeta^{k+1} r}(s_1, y_1)\right) \geq \frac{\lambda}{\alpha} - \frac{1}{\alpha}\frac{1}{\delta} C^m \zeta^{-km} - \frac{1}{\delta} C^m \zeta^{-(k+1)m}.$$

By (3.8.5)

$$\begin{aligned}\operatorname{Osc}\left(u, D_{\zeta^{k+1} r}(s_1, y_1)\right) \geq & \frac{1}{\alpha} \operatorname{Osc}\left(u, D_{\frac{1}{C}\zeta^{k+} r}(s_1, y_1)\right) \\ \geq & \frac{\lambda}{\alpha^2} - \frac{1}{\alpha^2}\frac{1}{\delta} C^m \zeta^{-km} - \frac{1}{\alpha}\frac{1}{\delta} C^m \zeta^{-(k+1)m}.\end{aligned}$$

and hence there is $(t_2, x_2) \in D_{\zeta^{k+1} r}(s_1, y_1)$ such that

$$u(t_2, x_2) \geq \frac{\lambda}{\alpha^2} - \frac{1}{\alpha^2}\frac{1}{\delta} C^m \zeta^{-km} - \frac{1}{\alpha}\frac{1}{\delta} C^m \zeta^{-(k+1)m}.$$

If $\frac{1}{C}\zeta^{k+2} r \leq 1$ then we stop there. If not we continue and if $\nu \in \mathbb{N}$ is such that

$$\frac{1}{C}\zeta^{k+\nu} r > 1 \geq \frac{1}{C}\zeta^{k+\nu+1} r, \tag{3.8.9}$$

then we obtain, in this way, finite sequences of points

$$(t_0, x_0), (t_1, x_1), ..., (t_\nu, x_\nu) \quad \text{and} \quad (s_0, y_0), (s_1, y_1), ..., (s_{\nu-1}, y_{\nu-1})$$

such that

$$\begin{gathered}(t_\mu, x_\mu) \in D_{\zeta^{k+\mu} r}(s_{\mu-1}, y_{\mu-1}), \\ D_{\frac{1}{C}\zeta^{k+\mu+1} r}(s_\mu, y_\mu) \subseteq D_{\zeta^{k+\mu} r}(s_{\mu-1}, y_{\mu-1})\end{gathered}$$

and

$$\begin{aligned}u(t_\mu, x_\mu) \geq \frac{\lambda}{\alpha^\mu} & - \frac{1}{\alpha^\mu}\frac{1}{\delta} C^m \zeta^{-km} - \frac{1}{\alpha^{\mu-1}}\frac{1}{\delta} C^m \zeta^{-(k+1)m} \\ & - \frac{1}{\alpha^{\mu-2}}\frac{1}{\delta} C^m \zeta^{-(k+2)m} - ... - \frac{1}{\alpha}\frac{1}{\delta} C^m \zeta^{-(k+\mu-1)m}\end{aligned} \tag{3.8.10}$$

for all $\mu = 1, 2, ..., \nu$.

Let us observe that by (3.8.7)

$$\sum_{\mu\geq 0}\zeta^{k+\mu}r=\zeta^k r\sum_{\mu\geq 0}\zeta^{\mu}=\zeta^k\frac{1}{1-\zeta}r\leq\frac{1}{4}r$$

and hence we have indeed that

$$\begin{aligned}(t_\mu,x_\mu)\in D_{\zeta^{k+\mu}r}(s_{\mu-1},y_{\mu-1})\subseteq&(\frac{7}{16}r^2,r^2)\times B_{\frac{3}{4}r}(e)\\ \subseteq&(1,r^2)\times B_r(e)\end{aligned}$$

for all $\mu=1,2,...,\nu$.

By (3.8.10)

$$\begin{aligned}u(t_\nu,x_\nu)\geq&\frac{\lambda}{\alpha^\nu}-\frac{1}{\alpha^\nu}\frac{1}{\delta}C^n\zeta^{-km}-\frac{1}{\alpha^{\nu-1}}\frac{1}{\delta}C^m\zeta^{-(k+1)m}-...-\frac{1}{\alpha}\frac{1}{\delta}C^m\zeta^{-(k+\nu-1)m}\\ =&\frac{\lambda}{\alpha^\nu}-\frac{1}{\alpha^\nu}\frac{1}{\delta}C^n\zeta^{-km}-\frac{1}{\alpha^\nu}\frac{1}{\delta}C^m\zeta^{-km}\alpha\zeta^{-m}-...-\frac{1}{\alpha^\nu}\frac{1}{\delta}C^m\zeta^{-km}\left(\alpha\zeta^{-m}\right)^{\nu-1}\\ =&\frac{1}{\alpha^\nu}\left[\lambda-\frac{1}{\delta}C^n\zeta^{-km}\left(1+\alpha\zeta^{-m}+\left(\alpha\zeta^{-m}\right)^2+...+\left(\alpha\zeta^{-m}\right)^{\nu-1}\right)\right]\\ \geq&\frac{1}{\alpha^\nu}\left[\lambda-\frac{1}{\delta}C^n\zeta^{-km}\left(1+\alpha\zeta^{-m}+\left(\alpha\zeta^{-m}\right)^2+...\right)\right]\\ =&\frac{1}{\alpha^\nu}\left(\lambda-\frac{1}{\delta}C^n\zeta^{-km}\frac{1}{1-\alpha\zeta^{-m}}\right).\end{aligned}$$

But, by (3.8.9)

$$\frac{1}{C}\zeta^{k+\nu+1}r\leq 1$$

and hence

$$(k+\nu+1)\log\zeta+\log r\leq\log C,$$

or

$$\nu\log\zeta\leq\log C-(k+1)\log\zeta-\log r,$$

or

$$\nu\geq\frac{\log C}{\log\zeta}-(k+1)-\frac{\log r}{\log\zeta}.$$

It follows that

$$\begin{aligned}\frac{1}{\alpha^\nu}=&e^{-\nu\log\alpha}\\ \geq&e^{\log\alpha\log r/\log\zeta}e^{-(k+1)\log\alpha}e^{-\log\alpha\log C/\log\zeta}\\ =&r^{\log\alpha/\log\zeta}\alpha^{-(k+1)}\alpha^{-\log C/\log\zeta}.\end{aligned}$$

So we have that

$$\begin{aligned}u(t_\nu,x_\nu)\geq&\frac{1}{\alpha^\nu}\left(\lambda-\frac{1}{\delta}C^n\zeta^{-km}\frac{1}{1-\alpha\zeta^{-m}}\right)\\ \geq&\frac{1}{\alpha^\nu}\frac{1}{\delta}\alpha^{\log C/\log\zeta}\geq r^{\log\alpha/\log\zeta}\alpha^{-(k+1)}\alpha^{-\log C/\log\zeta}\frac{1}{\delta}\alpha^{\log C/\log\zeta}\\ =&\frac{1}{\delta}r^{\log\alpha/\log\zeta}\alpha^{-(k+1)}>\frac{1}{\delta}r^{\log\alpha/\log\zeta}>\frac{1}{\delta}r^m\end{aligned}$$

which contradicts (3.8.3).

This proves (3.8.8) and the theorem follows.

4. Hölder continuity

Let us denote by ∂_1 and $\partial_z, z \in G$ respectively the difference operators

$$\partial_1 u(t,x) = u(t+1,x) - u(t,x) \quad \text{and} \quad \partial_z u(t,x) = u(t,xz) - u(t,x).$$

A repeated use of proposition 1.6.5, yields the following:

THEOREM 4.1. *There is $\gamma \in (0,1]$ and $c > 1$ such that for all $r \geq 2$ and every function u satisfying $\left(\frac{\partial}{\partial t} + L\right) u = 0$ in $(-r^2, 0] \times B_r(e)$*

$$|\partial_z u(t,x)| \leq c r^{-\gamma} \|u\|_\infty \tag{4.1}$$

for all $-1 \leq t \leq 0$ and $x, z \in B_1(e)$.

Let us now observe that by the left invariance of L, if u is a heat function then $v(s,y) = u(t+s, xy) - u(s,x)$ is also a heat function. So, combining (4.1) with (3.1) and the local Harnack inequality (2.2.1) we have the following:

COROLLARY 4.2. *Let $Y \in \mathfrak{g}$ and $a > 1$. Then, there is $\gamma \in (0,1]$, $c, C > 0$ such that for all $r \geq 1$ and every positive function $u \geq 0$ satisfying $\left(\frac{\partial}{\partial t} + L\right) u = 0$ in $(0, ar^2] \times B_{Cr}(x)$*

$$|Yu(r,x)| \leq c r^{-\gamma} u(ar^2, x) \tag{4.2}$$

An immediate consequence of (4.2) is the following:

COROLLARY 4.3. *Every function $u \geq 0$ satisfying $Lu = 0$ on G is constant.*

Combining (4.2) and (1.6.1), we have the following estimate for the heat kernel $p_t(x,y)$ of L:

COROLLARY 4.4. *Let $Y \in \mathfrak{g}$. Then, there is $\gamma \in (0,1]$ such that for all $Y \in \mathfrak{g}$ there is $c > 0$ such that*

$$|Y p_t(x,y)| \leq c t^{-(D+\gamma)/2} \tag{4.3}$$

for all $t \geq 1$ and $x, y \in G$.

5. Nilpotent Lie groups

In this section we shall recall certain results on the algebraic structure and the geometry of a simply connected nilpotent Lie group N. Related references, for this and the next section are [**A1, CR, FS, NRS, Va, V1, VSC**].

5.1. The filtration of the Lie algebra. Let us denote by $\mathfrak{n}$ the Lie algebra of N, which we identify with the left invariant vector fields on N.

We set $\mathfrak{n}_1 = \mathfrak{n}$ and $\mathfrak{n}_{i+1} = [\mathfrak{n}_1, \mathfrak{n}_i]$, $i \geq 1$. Since $\mathfrak{n}$ is nilpotent, we have the filtration:

$$\mathfrak{n} = \mathfrak{n}_1 \supseteq \mathfrak{n}_2 \supseteq \ldots \supseteq \mathfrak{n}_m \supseteq \mathfrak{n}_{m+1} = \{0\}, \ \mathfrak{n}_m \neq \{0\}.$$

We consider linear subspaces $\mathfrak{a}_1, \ldots, \mathfrak{a}_m$ of $\mathfrak{n}$ such that

$$\mathfrak{n}_i = \mathfrak{a}_i \oplus \ldots \oplus \mathfrak{a}_m, \quad 1 \leq i \leq m.$$

We set

$$n_0 = 0,\ n_i = \dim(\mathfrak{a}_1 \oplus ... \oplus \mathfrak{a}_i),\ 1 \le i \le m$$
$$\sigma(j) = i,\ \textit{for}\ n_{i-1} < j \le n_i$$
$$q = n_m = \dim(\mathfrak{n}).$$

Notice that the homogeneous dimension D of N is given by

$$D = \sigma(1) + ... + \sigma(q).$$

We consider a basis $\{X_1, ..., X_q\}$ of $\mathfrak{n}$ such that $\{X_{n_{i-1}+1}, ..., X_{n_i}\}$ is a basis of $\mathfrak{a}_i,\ 1 \le i \le m$.

5.2. A family of group structures and the limit group. On the linear space $\mathfrak{n}$, we define a family of Lie brackets $[.,.]_\varepsilon, \varepsilon \ge 0$, by setting

$$[X_i, X_j]_\varepsilon = \sum_{0 \le \ell \le m-\sigma(i)-\sigma(j)} \varepsilon^\ell \operatorname{pr}_{\mathfrak{a}_{\sigma(i)+\sigma(j)+\ell}}[X_i, X_j].$$

Notice that

$$[X_i, X_j]_0 = \operatorname{pr}_{\mathfrak{a}_{\sigma(i)+\sigma(j)}}[X_i, X_j]$$

and that the Lie algebras $\mathfrak{n}_\varepsilon = (\mathfrak{n}, [.,.]_\varepsilon),\ \varepsilon \ge 0$ are nilpotent.

Using the exponential coordinates of the second kind (or Malcev coordinates)

$$\phi : \mathbb{R}^q \to N,\ \phi : x = (x_q, ..., x_1) \to \exp x_q X_q ... \exp x_1 X_1$$

we identify N, as a differential manifold, with $\mathbb{R}^q$.

We denote by $\tau_\varepsilon, \varepsilon > 0$ the family of dilations

$$\tau_\varepsilon : (x_q, ..., x_1) \to (\varepsilon^{\sigma(q)} x_q, ..., \varepsilon^{\sigma(1)} x_1).$$

We define a family of group products $*_\varepsilon, \varepsilon > 0$, by setting

$$x *_\varepsilon y = \tau_\varepsilon \left[(\tau_{\varepsilon^{-1}} x)(\tau_{\varepsilon^{-1}} y) \right].$$

We also set

$$x *_0 y = \lim_{\varepsilon \to 0} x *_\varepsilon y.$$

In this way, we obtain a family of nilpotent Lie groups $N_\varepsilon = (N, *_\varepsilon), \varepsilon \ge 0$.

Note that the Lie algebra of N_ε is isomorphic to $\mathfrak{n}_\varepsilon$.

We shall identify $\mathfrak{n}_\varepsilon$ with the $*_\varepsilon$-left invariant vector fields.

If $X \in \mathfrak{n}$ is a left invariant vector field on N, then we shall denote by X_ε tha $*_\varepsilon$-left invariant vector field satisfying $X_\varepsilon(e) = X(e)$.

In particular we shall denote by $X_{\varepsilon i}$ the left invariant vector fields on N_ε satisfying $X_{\varepsilon i}(0) = X_i(0),\ 1 \le i \le q$.

Note that

$$X_{\varepsilon i} = \frac{1}{\varepsilon^{\sigma(i)}} d\tau_\varepsilon(X_i), \quad 1 \le i \le q. \tag{5.2.1}$$

Note also that N_0 is a stratified nilpotent Lie group (cf. [**FS**]) and that

$$X_{0i} = \frac{1}{\varepsilon^{\sigma(i)}} d\tau_\varepsilon(X_{0i}) \quad 1 \le i \le q. \tag{5.2.2}$$

DEFINITION. The above defined stratified nilpotent Lie group N_0 will be called limit group (at infinity) of N.

5.3. The vector fields $X_\varepsilon, 0 \le \varepsilon \le 1$ expressed in terms of the exponential coordinates. Having identified N with $\mathbb{R}^q$ using the exponential coordinates of the second kind, we shall give now an expression of the left invariant vectors fields of N as vector fields on $\mathbb{R}^q$. We shall need some more notation.

If $X = a_1X_1 + ... + a_qX_q$ then we shall set $\mathrm{pr}_i(X) = a_i, i = 1, ..., q$.

We denote by $\overline{ad}\ X_i$ the linear transformations of $\mathfrak{n}$ defined by

$$\overline{\mathrm{ad}}(X_i)X_j = 0, \quad \text{for } i \ge j \text{ and } \overline{\mathrm{ad}}(X_i)X_j = \mathrm{ad}(X_i)X_j, \text{ for } i < j.$$

LEMMA 5.3.1 (cf. [A1]). *Let X be a left invariant vector field on N. Then $X(x) = a_q(x)\frac{\partial}{\partial x_q} + ... + a_1(x)\frac{\partial}{\partial x_1}$ with*

$$\begin{aligned} a_i(x) &= \mathrm{pr}_i\left[e^{x_{i-1}\overline{\mathrm{ad}}X_{i-1}}...e^{x_1\overline{\mathrm{ad}}X_1}(X)\right] \\ &= \mathrm{pr}_i\Bigg[\sum_{\lambda_1\sigma(1)+...+\lambda_{i-1}\sigma(i-1)\le\sigma(i)-1} \frac{1}{\lambda_1!}\cdots\frac{1}{\lambda_{i-1}!}x_1^{\lambda_1}...x_{i-1}^{\lambda_{i-1}} \\ &\qquad (\overline{\mathrm{ad}}X_{i-1})^{\lambda_{i-1}}...(\overline{\mathrm{ad}}X_1)^{\lambda_1}(X)\Bigg]. \end{aligned} \tag{5.3.1}$$

COROLLARY 5.3.2. *Let X be a left invariant vector field on N and denote by X_ε the $*_\varepsilon$-left invariant vector field satisfying $X_\varepsilon(0) = X(0), \varepsilon \ge 0$. Then*

$$X_\varepsilon(x) = a_{\varepsilon q}(x)\frac{\partial}{\partial x_q} + ... + a_{\varepsilon 1}(x)\frac{\partial}{\partial x_1}, \tag{5.3.2}$$

where

$$\begin{aligned} a_{\varepsilon i}(x) = \mathrm{pr}_i\Bigg[&\sum_{\lambda_1\sigma(1)+...+\lambda_{i-1}\sigma(i-1)\le\sigma(i)-1} \varepsilon^{\sigma(i)-1-\lambda_1\sigma(1)-...-\lambda_{i-1}\sigma(i-1)} \\ &\frac{1}{\lambda_1!}\cdots\frac{1}{\lambda_{i-1}!}x_1^{\lambda_1}...x_{i-1}^{\lambda_{i-1}}(\overline{\mathrm{ad}}X_{i-1})^{\lambda_{i-1}}...(\overline{\mathrm{ad}}X_1)^{\lambda_1}(X)\Bigg]. \end{aligned}$$

It follows from (5.3.3) that

$$X_\varepsilon - X_0 = \varepsilon\left[b_{\varepsilon q}(x)\frac{\partial}{\partial x_q} + ... + b_{\varepsilon n_2+1}(x)\frac{\partial}{\partial x_{n_2+1}}\right], \tag{5.3.3}$$

where

$$\begin{aligned} b_{\varepsilon i}(x) = \mathrm{pr}_i\Bigg[&\sum_{\lambda_1\sigma(1)+...+\lambda_{i-1}\sigma(i-1)\le\sigma(i)-2} \varepsilon^{\sigma(i)-2-\lambda_1\sigma(1)-...-\lambda_{i-1}\sigma(i-1)} \\ &\frac{1}{\lambda_1!}\cdots\frac{1}{\lambda_{i-1}!}x_1^{\lambda_1}...x_{i-1}^{\lambda_{i-1}}(\overline{\mathrm{ad}}X_{i-1})^{\lambda_{i-1}}...(\overline{\mathrm{ad}}X_1)^{\lambda_1}(X)\Bigg]. \end{aligned}$$

5.4. Gradients. Let us set, for $f \in C^\infty$ and $\ell \in \mathbb{N}$,

$$\nabla^\ell f(x) = \sum_{\substack{a \le \ell \\ \sigma(i_1)+\ldots+\sigma(i_a) \ge \ell}} |\frac{\partial}{\partial x_{i_1}} \cdots \frac{\partial}{\partial x_{i_a}} f(x)|,$$

$$\nabla^\ell_X f(x) = \sum_{\substack{a \le \ell \\ \sigma(i_1)+\ldots+\sigma(i_a) \ge \ell}} |X_{i_1} \ldots X_{i_a} f(x)|$$

and more generally

$$\nabla^\ell_{\varepsilon X} f(x) = \sum_{\substack{a \le \ell \\ \sigma(i_1)+\ldots+\sigma(i_a) \ge \ell}} |X_{\varepsilon i_1} \ldots X_{\varepsilon i_a} f(x)|.$$

The following lemma is an immediate consequence of (5.3.1) and (5.3.2).

LEMMA 5.4.1. *There is $c > 0$ and $N \in \mathbb{N}$ such that*

$$\frac{1}{c(1+|x|)^N} \nabla^\ell f(x) \le \nabla^\ell_{\varepsilon X} f(x) \le c(1+|x|)^N \nabla^\ell f(x) \tag{5.4.1}$$

for all $x \in N$ and $0 \le \varepsilon \le 1$.

6. Sub-Laplacians on nilpotent Lie groups

Let $L = -\left(E_1^2 + \ldots + E_p^2\right) + E_0$ be a centered sub-Laplacian on a simply connected nilpotent Lie group N. Let $\mathfrak{n}$ be the Lie algebra of N. Then the assumption that L is centered implies that $E_0 \in [\mathfrak{n}, \mathfrak{n}]$.

Let $\{X_1, \ldots, X_q\}$ be the basis of $\mathfrak{n}$ introduced in section 5. Since the vector fields E_i are linear combinations of the vector fields X_i, L can also be written as

$$L = -\sum_{1 \le i,j \le q} a_{ij} X_i X_j + \sum_{n_1 < i \le q} a_i X_i. \tag{6.1}$$

Note that $a_{ij} = a_{ji}, 1 \le i, j \le q$. Also, since the vector fields $E_1, \ldots, E_p$ satisfy Hörmander's condition (i.e. they generate, together with their successive Lie brackets, the Lie algebra $\mathfrak{n}$), the $(n_1 \times n_1)$ matrix $B = (b_{ij})$ with entries $b_{ij} = a_{ij}, 1 \le i, j \le n_1$ is positive definite.

6.1. The rescaled sub-Laplacians L_ε and the limit sub-Laplacian L_0. Let $d\tau_\varepsilon, \varepsilon > 0$ be the family of dilations introduced in section 5.2 and let

$$L_\varepsilon = \frac{1}{\varepsilon^2} d\tau_\varepsilon(L).$$

Then by (5.2.1)

$$L_\varepsilon = -\sum_{1 \le i,j \le q} \varepsilon^{\sigma(i)+\sigma(j)-2} a_{ij} X_{\varepsilon i} X_{\varepsilon j} + \sum_{n_1 < i \le q} \varepsilon^{\sigma(i)-2} a_i X_{\varepsilon i}. \tag{6.1.1}$$

The sub-Laplacians L_ε are $*_\varepsilon$-left invariant. If we let $\varepsilon \to 0$, then we get the $*_0$-left invariant sub-Laplacian

$$L_0 = -\sum_{1 \le i,j \le n_1} a_{ij} X_{0i} X_{0j} + \sum_{n_1 < i \le n_2} a_i X_{0i}. \tag{6.1.2}$$

Let us observe that L_0 is dilation invariant, i.e. that

$$L_0 = \frac{1}{\varepsilon^2} d\tau_\varepsilon(L_0), \qquad \varepsilon > 0. \tag{6.1.3}$$

DEFINITION. The above defined sub-Laplacian L_0 is called limit sub-Laplacian (at infinity) associated to L.

6.2. Comparison of the control diastances. Let $d(x,y)$ and $d_\varepsilon(x,y)$ be the control distances associated to the sub-Laplacians L and $L_\varepsilon, 0 \le \varepsilon \le 1$ respectively. Note that these distances are left invariant, i.e. $d_\varepsilon(z *_\varepsilon x, z *_\varepsilon y) = d_\varepsilon(x,y),\ x,y,z \in N$.

Let $B_r^\varepsilon(x) = \{y \in N : d_\varepsilon(x,y) < r\}$ denote the balls of radius r and centered at $x \in N$.

By (5.2.1),

$$\begin{aligned} d(x,y) &= \frac{1}{\varepsilon} d_\varepsilon(\tau_\varepsilon x, \tau_\varepsilon y), \quad \varepsilon > 0 \ \ x,y \in N \\ d_0(x,y) &= \frac{1}{\varepsilon} d_0(\tau_\varepsilon x, \tau_\varepsilon y), \quad x,y \in N. \end{aligned} \tag{6.2.1}$$

Let us fix a compact neighborhood U of the identity element e of N and let us define $|.|_N$ as in (1.1). Then, it follows from (5.3.2) and (5.3.3) that $\cap_{0\le\varepsilon\le1} B_1^\varepsilon(e)$ is a neighborhood of e. So, there is a constant $c > 0$ such that

$$B_{1/c}^\varepsilon(e) \subseteq U \subseteq B_c^\varepsilon(e), \quad 0 \le \varepsilon \le 1.$$

Let $U^{*_\varepsilon n} = \{g_1 *_\varepsilon g_2 *_\varepsilon \ldots *_\varepsilon g_n : g_1, g_2, \ldots, g_n \in U\}$. Then

$$B_{n/c}^\varepsilon(e) \subseteq U^{*_\varepsilon n} \subseteq B_{cn}^\varepsilon(e), \quad n \in \mathbb{N}, \ 0 \le \varepsilon \le 1. \tag{6.2.2}$$

Now, let $c > 0$ be a constant satisfying

$$B_{1/c}^0(e) \subseteq B_1^\varepsilon(e) \subseteq B_c^0(e), \quad 1 \le \varepsilon \le 1.$$

Then, it follows from (6.2.1) that

$$B_{r/c}^0(e) \subseteq B_r^\varepsilon(e) \subseteq B_{cr}^0(e), \quad 0 \le \varepsilon \le 1, \quad r \ge 1. \tag{6.2.3}$$

It follows from (6.2.2) and (6.2.3) that there is a constant $c > 1$ such that

$$\frac{1}{c}|x|_N \le 1 + d_\varepsilon(e,x) \le c|x|_N, \quad 0 \le \varepsilon \le 1. \tag{6.2.4}$$

6.3. A Harnack inequality for dilation invariant sub-Laplacians. Let us observe that if the function $u(t,x)$ satisfies

$$\left(\frac{\partial}{\partial t} + L_0\right) u = 0 \quad \text{in} \quad (0,T) \times B_r^0(e)$$

then the function $u_\varepsilon(t,x) = u(t/\varepsilon^2, \tau_{1/\varepsilon} x)$ will satisfy

$$\left(\frac{\partial}{\partial t} + L_0\right) u_\varepsilon = 0 \quad \text{in} \quad (0, \varepsilon^2 T) \times B_{\varepsilon r}^0(e).$$

Combining this observation and (2.2.1) we have the following:

THEOREM 6.3.1. *Let $0 < a_1 \le a_2 < b_1 \le b_2 < 1$ and $\rho \in (0,1)$. Then, for all $k, n \in \mathbb{N}$ and there is a constant $c > 0$ for all $1 \le i_j \le q, 1 \le j \le n$ and every every positive solution u of $\left(\frac{\partial}{\partial t} + L_0\right) u = 0$ on $(0,t) \times B^0_{\sqrt{t}}(e)$, we have*

$$\text{(6.3.1)} \quad \sup\left\{ |\frac{\partial^k}{\partial t^k} X_{0i_1} ... X_{0i_n} u|; [a_1 t, a_2 t] \times B^0_{\rho\sqrt{t}}(e) \right\} \le c_{k,n} t^{-k-\left(\sigma(i_1)+...\sigma(i_n)\right)/2} \inf\left\{ u \ ; [b_1 t, b_2 t] \times B^0_{\rho\sqrt{t}}(e) \right\}.$$

6.4. The heat kernels. Let $p^\varepsilon_t(x,y)$ be the heat kernel of L_ε , $0 \le \varepsilon \le 1$ (and $p_t(x,y)$ the heat kernel of L). Then

$$\text{(6.4.1)} \qquad p_t(x,y) = \varepsilon^D p^\varepsilon_{\varepsilon^2 t}(\tau_\varepsilon(x), \tau_\varepsilon(y)), \quad \varepsilon > 0.$$

Since L_0 is dilation invariant, we have

$$\text{(6.4.2)} \qquad p^0_t(x,y) = \varepsilon^D p^0_{\varepsilon^2 t}(\tau_\varepsilon x, \tau_\varepsilon y), \quad \varepsilon > 0.$$

It follows that there is a constant $C_{L_0} > 0$ such that

$$\text{(6.4.3)} \qquad p^0_t(e,e) = C_{L_0} t^{-D/2}.$$

6.5. A Gaussian estimate for $p^0_t(x,y)$. The heat kernel $p^0_t(x,y)$ of L_0 satisfies the following Gaussian estimate:

THEOREM 6.5.1. *There is a constant $c > 0$ such that*

$$\text{(6.5.1)} \qquad p^0_t(x,y) \le c t^{-D/2} \exp\left(-\frac{d_0(x,y)^2}{ct}\right), \quad x, y \in N, t > 0.$$

Since L_0 is dilation invariant, to prove (6.5.1), it is enough to prove that there is $c > 0$ such that

$$p^0_1(x,e) \le c \exp\left(-\frac{|x|^2_N}{c}\right), \quad x \in N.$$

This is a wellknown result (see for example [**V8,** appendix A4]). The proof given in section 9 also works in this case (just take $k = 1$ and leave everything else the same).

Combining (6.5.1) with (6.3.1) we have the following:

COROLLARY 6.5.2. *For all $0 \le k \in \mathbb{Z}$ and $0 \le n \in \mathbb{Z}$ there is a constant $c > 0$ such that*

$$\text{(6.5.2)} \quad |\frac{\partial^k}{\partial t^k} X_{0i_1} ... X_{0i_n} p^0_t(x,y)| \le c t^{-\left(D+2k+\sigma(i_1)+...+\sigma(i_n)\right)/2} \exp\left(-\frac{d_0(x,y)^2}{ct}\right),$$

for all $t > 0$, $x, y \in N$ and $1 \le i_j \le q$, $1 \le j \le n$.

7. A function which grows linearly

The goal of this section is to construct, on a simply connected nilpotent Lie group N, a positive function $\rho(x)$ which can serve as a smooth substitute for $|x|_N$. This function will be a convenient power of the Green function of a dilation invariant sub-Laplacian L_0 on N.

We shall use the notation of the previous section.

PROPOSITION 7.1. *Let N be a simply connected nilpotent Lie group. Then, there is a function $\rho(x) \in C^\infty(N)$ with the following properties: For all $n \in \mathbb{N}$ there is a constant $c \geq 1$ such that for all $x \in N$, all $1 \leq i_j \leq q, 1 \leq j \leq n$ and all $0 \leq \varepsilon \leq 1$,*

$$\begin{aligned} &\rho(x) \geq 0, \quad x \in N \\ &\frac{1}{c}\ |x|_N \leq \rho(x) \leq\ c\ |x|_N \quad \textit{for } |x| \geq 2 \qquad (7.1)\\ &|X_{\varepsilon i_1} \ldots X_{\varepsilon i_n} \rho(x)| \leq \frac{c}{|x|_N^{\sigma(i_1)+\ldots+\sigma(i_n)-1}}, \qquad \textit{for } |x| \geq 2. \end{aligned}$$

PROOF. If the homogeneous dimension D of N is $D \leq 2$, then N is isomorphic either to $\mathbb{R}$ or to $\mathbb{R}^2$ and then we can take as ρ the Euclidean norm. So let us assume that $D > 2$. Let L be a a left invariant sub-Laplacian on N and let L_0 the associated dilation invariant sub-Laplacian. Then the Green function of L_0 is given by

$$G_{L_0}(x, y) = \int_0^\infty p_t^0(x, y) dt.$$

Let $G_{L_0}(x) = G_{L_0}(x, e)$. Then by the dilation invariance of L_0,

$$G_{L_0}(x) = \varepsilon^{D-2} G_{L_0}(\tau_\varepsilon x), \quad x \in N, \varepsilon > 0. \qquad (7.2)$$

We have that $L_0 G_{L_0}(x) = 0$, in $N \setminus \{e\}$. So, by taking $\varepsilon = 1/|x|_N$ and using (6.3.1) we have that there is $c > 0$ such that

$$\frac{1}{c}\ \frac{1}{|x|_N^{D-2}} \leq G_{L_0}(x) \leq\ c\ \frac{1}{|x|_N^{D-2}}, \quad |x|_N \geq 2. \qquad (7.3)$$

Furthermore, for all $n \in \mathbb{N}$ there is $c > 0$ such that

$$|X_{0i_1} X_{0i_2} \ldots X_{0i_n} G_{L_0}(x)| \leq c, \quad 1 \leq |x|_N \leq 2.$$

Actually, by (5.3.2) and the local Harnack inequality (2.2.1) we also have

$$|X_{\varepsilon i_1} X_{\varepsilon i_2} \ldots X_{\varepsilon i_n} G_{L_0}(x)| \leq c, \quad 1 \leq |x|_N \leq 2, \quad 0 \leq \varepsilon \leq 1.$$

By rescaling, we get that there is $c > 0$ such that

$$|X_{\varepsilon i_1} X_{\varepsilon i_2} \ldots X_{\varepsilon i_n} G_{L_0}(x)| \leq \frac{c}{|x|_N^{D-2+\sigma(i_1)+\ldots+\sigma(i_n)}}, \quad |x|_N \geq 2, \quad 0 \leq \varepsilon \leq 1. \qquad (7.4)$$

The function

$$\rho(x) = \phi(x) \left(G_{L_0}(x)\right)^{-1/(D-2)}$$

satisfies (7.1).

8. Proof of propositions 1.6.3 and 1.6.4 in the case of nilpotent Lie groups

The goal of this section is prove propositions 1.6.3 and 1.6.4 when L is a centered left invariant sub-Laplacian on a simply connected Lie group N.

Let L_ε, $0 \le \varepsilon \le 1$ be the rescaled sub-Laplacians introduced in section 6 and let $p_t^\varepsilon(x,y)$, $0 \le \varepsilon \le 1$ their heat kernels.

If $K(x,y)$ and $S(x,y)$ are two kernels then we shall denote by KS their product $KS(x,y) = \int K(x,z)S(z,y)dy$.

The proof of propositions 1.6.3 and 1.6.4 is based on the following:

LEMMA 8.1. *For all $r, t_1, T > 0$, there is a constant $c = c(r, t_1, T) > 0$ such that*

$$|\,(p_t^\varepsilon - p_t^0)\, p_T^0(x,y)| \le c\, \varepsilon \tag{8.1}$$

for all $x \in N, |y|_N \le r,\ 0 \le t \le t_1,$ and $0 \le \varepsilon \le 1$.

PROOF. Let us fix $y \in N$ and consider the function $u_\varepsilon(t,x) = (p_t^\varepsilon - p_t^0)p_T^0(x,y)$. Let us set

$$v_\varepsilon(t,x) = (\frac{\partial}{\partial t} + L_\varepsilon)u_\varepsilon(t,x).$$

Then $v_\varepsilon(t,x) == (L_\varepsilon - L_0)p_{t+T}^0(x,y)$. Since $u_\varepsilon(0,x) = 0,\ x \in N$, we have

$$u_\varepsilon(t,x) = \int_0^t \int p_{t-s}^\varepsilon(x,z)v_\varepsilon(s,z)dzds.$$

Now, by (5.3.3) and (5.4.1), there is $k \in \mathbb{N}$ such that

$$\begin{aligned} |u_\varepsilon(t,x)| \le & \int_0^t \int p_{t-s}^\varepsilon(x,z)|v_\varepsilon(s,z)|dzds \\ = & \int_0^t \int p_{t-s}^\varepsilon(x,z)|(L_\varepsilon - L_0)p_{s+T}^0(z,y)|dzds \\ \le & c\,\varepsilon \int_0^t \int p_{t-s}^\varepsilon(x,z)(1+|z|_N)^k|\nabla_{0X}^2 p_{s+T}^0(z,y)|dzds \\ \le & c\,\varepsilon \int_0^t \|p_{t-s}^\varepsilon(x,.)\|_1 \|(1+|z|_N)^k \nabla_{0X}^2 p_{s+T}^0(z,y)\|_\infty ds. \end{aligned}$$

Since by (6.5.2)

$$\sup\left\{\|(1+|z|_N)^k \nabla_{0X}^2 p_{s+T}^0(z,y)\|_\infty : |y|_N \le r,\ 0 \le s \le t_1\right\} < \infty,$$

the lemma follows.

8.1. Proof of proposition 1.6.3. By the local Harnack inequality (2.2.1), (1.6.3) holds if we assume that $1 \le n \le n_0$, for any $n_0 \in \mathbb{N}$. So it is enough to prove that there is $n_0 \in \mathbb{N}$ such that (1.6.3) is satisfied for all $n \ge n_0$. This follows from the following lemma (we use the notation of section 6.2):

LEMMA 8.1.1. *Let $b > 0$ such that*

$$B_b^0(e) \subseteq B_1^\varepsilon(e), \quad 0 \leq \varepsilon \leq 1.$$

Then, for all $a > 1$ there is $\varepsilon_0 \in (0,1)$ and $\delta > 0$ such that

$$\int_{B_b^0(e)} p_t^\varepsilon(x,y)dy > \delta$$

for all $\varepsilon \in [0, \varepsilon_0]$ and $(t,x) \in (a^{-2}, a^2) \times B_a^0(e)$.

PROOF. Let $a > 1$ and $T \in (0,1)$ and set

$$m_a = \inf\{p_t^0(x,e) : x \in B_a^0(e), a^{-2} \leq t \leq a^2 + 1\}$$

and

$$M_T = \sup\{p_T^0(x,e) : x \notin B_b^0(e)\}.$$

Then, by (8.1) there is $c = c(a,T) > 0$ such that

$$\begin{aligned}
\int_{B_b^0(e)} p_t^\varepsilon(x,y)dy \geq & \frac{1}{\|p_T^0\|_\infty} \int_{B_b^0(e)} p_t^\varepsilon(x,y)p_T^0(y,e)dy \\
= & \frac{1}{\|p_T^0\|_\infty}\left[\int p_t^\varepsilon(x,y)p_T^0(y,e)dy - \int_{\{y \notin B_b^0(e)\}} p_t^\varepsilon(x,y)p_T^0(y,e)dy\right] \\
\geq & \frac{1}{\|p_T^0\|_\infty}\left[\int p_t^\varepsilon(x,y)p_T^0(y,e)dy - M_T\int p_t^\varepsilon(x,y)dy\right] \\
\geq & \frac{1}{\|p_T^0\|_\infty}\left[\int p_t^\varepsilon(x,y)p_T^0(y,e)dy - M_T\right] \\
= & \frac{1}{\|p_T^0\|_\infty}\left[\int p_t^0(x,y)p_T^0(y,e)dy\right. \\
& \left. + \int \left(p_t^\varepsilon(x,y) - p_t^0(x,y)\right)p_T^0(y,e)dy - M_T\right] \\
\geq & \frac{1}{\|p_T^0\|_\infty}\left[p_{t+T}^0(x,e) - |(p_t^\varepsilon - p_t^0)p_T^0(x,e))| - M_T\right] \\
\geq & \frac{1}{\|p_T^0\|_\infty}\left[p_{t+T}^0(x,e) - c\,\varepsilon - M_T\right] \\
\geq & \frac{1}{\|p_T^0\|_\infty}\left[m_a - c\,\varepsilon - M_T\right]
\end{aligned}$$

for all $(t,x) \in (a^{-2}, a^2) \times B_a^0(e)$.

So, if we chose first $T \in (0.1]$ small enough so that $M_T < m_a/4$ and then $\varepsilon_0 \in (0,1)$ so that $c\varepsilon_0 < m_a/4$, we have

$$\int_{B_b^0(e)} p_t^\varepsilon(x,y)dy \geq \frac{1}{\|p_T^0\|_\infty}\frac{m_a}{2}$$

for all $(t,x) \in (a^{-2}, a^2) \times B_a^0(e)$ and $0 \leq \varepsilon \leq \varepsilon_0$ and the lemma follows.

8.2. Proof of proposition 1.6.4. By the semigroup property $p_{t+s} = p_t p_s$ and the fact that

$$\int_U p_t(e,y)dy \to 1, \quad \text{as} \quad t \to 0,$$

for every neighborhood U of e, we have that (1.6.4) holds if we assume that $1 \leq n \leq n_0$, for any $n_0 \in \mathbb{N}$. So it is enough to prove that there is $n_0 \in \mathbb{N}$ such that (1.6.4) is satisfied for all $n \geq n_0$. This follows from the following lemma:

LEMMA 8.2.1. *For all $\delta > 0$ there is $c_\delta > 1$ and $\varepsilon_0 \in (0,1]$ such that*

$$\int_{\{y \in G: d_0(e,y) \geq c_\delta\}} p_1^\varepsilon(e,y)dy < \delta \tag{8.2.1}$$

for all $0 \leq \varepsilon \leq \varepsilon_0$.

PROOF. Let $r_1 > r_2 \geq 1$ and set

$$M = \sup\left\{\int_{B^0_{r_2}(e)} p_1^0(y,z)dz : y \notin B^0_{r_1}(e)\right\}.$$

Then by (8.1) there is $c = c(r_2) > 0$ such that,

$$\sup\left\{|(p_1^\varepsilon - p_1^0)p_1^0(e,z)| : z \in B^0_{r_2}(e)\right\} \leq c'\ \varepsilon, \quad 0 \leq \varepsilon \leq 1.$$

We have

$$\begin{aligned}
\int_{B^0_{r_1}(e)} p_1^\varepsilon(e,y)dy &\geq \int_{B^0_{r_1}(e)}\int_{B^0_{r_2}(e)} p_1^\varepsilon(e,y)p_1^0(y,z)dydz \\
&= \int\int_{B^0_{r_2}(e)} p_1^\varepsilon(e,y)p_1^0(y,z)dydz \\
&\quad - \int_{\{y \notin B^0_{r_1}(e)\}}\int_{B^0_{r_2}(e)} p_1^\varepsilon(e,y)p_1^0(y,z)dydz \\
&\geq \int\int_{B^0_{r_2}(e)} p_1^\varepsilon(e,y)p_1^0(y,z)dydz - M \qquad (8.2.2)\\
&= \int\int_{B^0_{r_2}(e)} p_1^0(e,y)p_1^0(y,z)dydz \\
&\quad + \int\int_{B^0_{r_2}(e)} \left(p_1^\varepsilon(e,y) - p_1^0(e,y)\right) p_1^0(y,z)dydz - M \\
&\geq \int_{B^0_{r_2}(e)} p_2^0(e,z)dz - c\ \varepsilon\ |B^0_{r_2}(e)| - M.
\end{aligned}$$

Now let $\delta > 0$ and let us chose $r_2 \geq 1$ such that

$$\int_{B^0_{r_2}(e)} p_2^0(e,z)dz \geq 1 - \frac{\delta}{3}.$$

Next, let us chose $r_1 > r_2$ such that $M \leq \frac{\delta}{3}$ and $\varepsilon_0 \in (0,1)$ such that

$$c\ \varepsilon_0\ |B^0_{r_2}(e)| \leq \frac{\delta}{3}.$$

Then, it follows from (8.2.2) that

$$\int_{B^0_{r_1}(e)} p^\varepsilon_1(e,y)dy \geq 1-\delta$$

and the lemma follows.

9. Proof of the Gaussian estimate in the case of nilpotent Lie groups

The goal of this section is to prove Gaussian estimate (1.8.1) when $p_t(x,y)$ is the heat kernel of a centered left invariant sub-Laplacian L on a simply connected nilpotent Lie group N.

9.1. The functions ρ_k, $k \geq 1$. Let $\rho(x)$ be as in section 7 and let the family of dilations $\tau_\varepsilon, \varepsilon > 0$ be as in section 5.2.

Let U be the compact neighborhood of the identity element e of N and let $|.|_N$ be defined as in (1.1). Then, by (6.2.3) there is $C > 1$ such that

$$\tau_{1/Cn} U^n \subseteq U, \quad n \in \mathbb{N}. \tag{9.1.1}$$

Let $\phi \in C^\infty(N)$, $\phi \geq 0$ such that

$$\phi(x) = \begin{cases} 0 & \text{for } |x|_N \leq 1 \\ 1 & \text{for } |x|_N \geq 4 \end{cases}$$

and set

$$\phi_k(x) = \phi(\tau_{1/C\sqrt{k}}x), \quad k \geq 1.$$

Then,

$$\phi_k(x) = 0, \quad |x|_N \leq \sqrt{k} \tag{9.1.2}$$

and there is a constant $\zeta > 0$ such that

$$\phi_k(x) = 1, \quad |x|_N \geq \zeta\sqrt{k}. \tag{9.1.3}$$

Also, if $\{X_1, ..., X_q\}$ is the basis of $\mathfrak{n}$ introduced in section 5.1, then for every $n \in N$ there is a constant $c > 0$

$$|X_{i_1}X_{i_2}...X_{i_n}\phi_k(x)| \leq ck^{-(\sigma(i_1)+...+\sigma(i_n))/2} \tag{9.1.4}$$

for all $x \in N$ and $1 \leq i_j \leq q, 1 \leq j \leq n$.

We set

$$\rho_k(x) = \phi_k(x)\rho(x).$$

In the next lemma we gather the properties of the functions $\rho_k(x)$ that we shall need in the proof of (1.8.1). These properties are immediate consequences of (7.1).

LEMMA 9.1.1. *For all $n \in \mathbb{N}$ there is a constant $c \geq 1$ such that for all $k \geq 1$ and all $1 \leq i_j \leq q, 1 \leq j \leq n$,*

$$\begin{aligned} &\rho_k(x) \geq 0, \quad x \in N \\ &\rho_k(x) = 0, \quad \text{for } |x|_N \leq \sqrt{k} \\ &\frac{1}{c}\,|x|_N \leq \rho_k(x) \leq \; c\,|x|_N \quad \text{for } |x| \geq \zeta\sqrt{k} \\ &|X_{i_1}...X_{i_n}\rho_k(x)| \leq \frac{c}{|x|_N^{\sigma(i_1)+...+\sigma(i_n)-1}}, \quad x \in N. \end{aligned} \tag{9.1.5}$$

9.2. The functions H_k, $k \geq 1$. For fixed constants $A > 0$ and $B > 0$ we consider the family of functions H_k, $k \geq 1$ defined by

$$H_k(t,x) = \exp\left(-\frac{\left(\rho_k(x) + B\sqrt{k}\right)^2}{A(k+t)} \right), \quad t \geq 0, \ x \in N.$$

LEMMA 9.2.1. *There are constants $A > 0$ and $B > 0$ such that*

$$\left(\frac{\partial}{\partial t} + L\right) H_k(t,x) > 0 \tag{9.2.1}$$

for all $k \geq 1$ and all $(t,x) \in [0,k] \times N$.

PROOF. Let us first recall that, with the notation of section 6, the sub-Laplacian L can also be written as

$$L = -\sum_{1 \leq i,j \leq q} a_{ij} X_i X_j + \sum_{n_1 < i \leq q} a_i X_i.$$

We have

$$\frac{\partial}{\partial t} H_k(t,x) = \frac{1}{A}\frac{1}{k+t} \frac{\left(\rho_k(x) + B\sqrt{k}\right)^2}{k+t} H_k(t,x).$$

Also, for all smooth vector fields X and Y, we have

$$\begin{aligned} YH_k(t,x) &= -\frac{1}{A}\frac{1}{k+t} 2\left(\rho_k(x) + B\sqrt{k}\right) Y\rho_k(x)\ H_k(t,x) \\ XYH_k(t,x) &= -\frac{1}{A}\frac{1}{k+t} 2X\rho_k(x) Y\rho_k(x)\ H_k(t,x) \\ &\quad -\frac{1}{A}\frac{1}{k+t} 2\left(\rho_k(x) + B\sqrt{k}\right) XY\rho_k(x)\ H_k(t,x) \\ &\quad +\frac{1}{A^2}\frac{1}{(k+t)^2} 4\left(\rho_k(x) + B\sqrt{k}\right)^2 X\rho_k(x) Y\rho(x)\ H_k(t,x). \end{aligned}$$

So

$$\begin{aligned} \left(\frac{\partial}{\partial t} + L\right) H_k(t,x) = &\frac{1}{A}\frac{1}{k+t}\Bigg[\frac{\left(\rho_k(x) + B\sqrt{k}\right)^2}{k+t} \\ &+ 2\sum_{1 \leq i,j \leq q} a_{ij} X_i\rho_k(x) X_j\rho_k(x) \\ &+ 2\left(\rho_k(x) + B\sqrt{k}\right) \sum_{1 \leq i,j \leq q} a_{ij} X_i X_j \rho_k(x) \\ &- 4\frac{1}{A}\frac{1}{k+t}\left(\rho_k(x) + B\sqrt{k}\right)^2 \sum_{1 \leq i,j \leq q} a_{ij} X_i\rho(x) X_j\rho_k(x) \\ &+ 2\left(\rho_k(x) + B\sqrt{k}\right) \sum_{n_1 < i \leq q} a_i X_i \rho_k(x) \Bigg] H_k(t,x). \end{aligned}$$

Case I: $|x|_N \leq \sqrt{k}$. Then

$$\left(\frac{\partial}{\partial t} + L\right) H_k(t,x) = \frac{1}{A}\frac{1}{k+t}\frac{B^2 k}{k+t} H_k(t,x) > 0.$$

Case II: $\sqrt{k} \le |x|_N \le \zeta\sqrt{k}$ and $0 \le t \le k$. Then by (9.1.5) there is $c > 0$ such that

$$\begin{aligned}\left(\frac{\partial}{\partial t}+L\right)H_k(t,x) \geq & \frac{1}{A}\frac{1}{k+t}\left[\frac{B^2}{2}-c-c\left(c\sqrt{k}+B\sqrt{k}\right)\frac{1}{\sqrt{k}}\right.\\ & \left.-c\frac{1}{A}\frac{1}{k}\left(c\sqrt{k}+B\sqrt{k}\right)^2-c\left(c\sqrt{k}+B\sqrt{k}\right)\frac{1}{\sqrt{k}}\right]H_k(t,x)\\ =&\frac{1}{A}\frac{1}{k+t}\left[\frac{B^2}{2}-c-c\,(c+B)-c\frac{1}{A}\,(c+B)^2-c\,(c+B)\right]H_k(t,x).\end{aligned}$$

So, by chosing B large enough, so that

$$\frac{B^2}{4} > c + c\,(c+B) + c\,(c+B)$$

and A large enough, so that

$$\frac{B^2}{4} > c\frac{1}{A}\,(c+B)^2$$

we have

$$\left(\frac{\partial}{\partial t}+L\right)H(t,x) > 0.$$

Case III: $|x|_N > \zeta\sqrt{k}$ and $0 \le t \le k$.
Then by (9.1.5) there are constants $c_1, c_2 > 0$ such that

$$\begin{aligned}\left(\frac{\partial}{\partial t}+L\right)H_k(t,x) \geq & \frac{1}{A}\frac{1}{k+t}\left[\frac{\left(c_1|x|_N+B\sqrt{k}\right)^2}{2k}-c_2-c_2\left(c_2|x|_N+B\sqrt{k}\right)\frac{1}{|x|_N}\right.\\ & \left.-c_2\frac{1}{A}\frac{1}{k}\left(c_2|x|_N+B\sqrt{k}\right)^2-c_2\left(c_2|x|_N+B\sqrt{k}\right)\frac{1}{|x|_N}\right]H_k(t,x)\\ =&\frac{1}{A}\frac{1}{k+t}\left[\frac{\left(c_1|x|_N+B\sqrt{k}\right)^2}{2k}-c_2-c_2^2-c_2B\frac{\sqrt{k}}{|x|_N}\right.\\ & \left.-c_2\frac{1}{A}\frac{\left(c_2|x|_N+B\sqrt{k}\right)^2}{k}-c_2^2-c_2B\frac{\sqrt{k}}{|x|_N}\right]H_k(t,x)\\ \geq&\frac{1}{A}\frac{1}{k+t}\left[\frac{\left(c_1|x|_N+B\sqrt{k}\right)^2}{2k}-c_2-c_2^2-c_2B\right.\\ & \left.-c_2\frac{1}{A}\frac{\left(c_2|x|_N+B\sqrt{k}\right)^2}{k}-c_2^2-c_2B\right]H_k(t,x).\end{aligned}$$

So, by chosing B large enough, so that

$$\frac{\left(c_1|x|_N+B\sqrt{k}\right)^2}{4k} > c_2+c_2^2+c_2+Bc_2^2+c_2B$$

and A large enough, so that

$$\frac{1}{4}\left(c_1|x|_N+B\sqrt{k}\right)^2 > c_2\frac{1}{A}\left(c_2|x|_N+B\sqrt{k}\right)^2$$

we have

$$\left(\frac{\partial}{\partial t}+L\right) H_k(t,x) > 0$$

for all $t \in [0,k]$ and $x \in N$ and the lemma follows.

9.3. Proof of theorem 1.8.1. It is enough to prove that there is $c > 0$ such that

$$p_t(x,e) \leq ct^{-D/2} \exp\left(-\frac{|x|_N^2}{ct}\right), \quad t \geq 1,\ x \in N. \tag{9.3.1}$$

Let us fix constants $A > 0$ and $B > 0$ such that the family of functions

$$H_k(t,x) = \exp\left(-\frac{\left(\rho_k(x) + B\sqrt{k}\right)^2}{A(k+t)}\right), \quad k \geq 1$$

satisfy (9.2.1). Let us also fix a function $f(x) \in C^\infty$ such that

$$\begin{aligned} &0 \leq f(x) \leq 1, && x \in N \\ &f(x) = 1, && |x|_N \leq \sqrt{k} \\ &f(x) = 0, && |x|_N \geq 2\sqrt{k}. \end{aligned}$$

We consider the function

$$u(t,x) = \int p_t(x,y) f(y) dt, \quad x \in N, t \geq 0.$$

Note that

$$\begin{aligned} &\left(\frac{\partial}{\partial t}+L\right) u(t,x) - 0, \quad x \in N, t > 0 \\ &u(0,x) = f(x), \quad x \in N. \end{aligned}$$

We fix a constant $C > 0$ such that

$$CH_k(0,x) > 1, \quad |x|_N \leq 3\sqrt{k}$$

and consider the function

$$F(t,x) = CH_k(t,x) - u(t,x).$$

Then $F(t,x)$ satisfies

$$\begin{aligned} &\left(\frac{\partial}{\partial t}+L\right) F(t,x) > 0, \quad x \in N, t \in (0,k] \\ &F(0,x) > 0, \quad x \in N. \end{aligned}$$

LEMMA 9.3.1. *The above defined function $F(t,x)$ satisfies*

$$F(t,x) \geq 0, \quad t \in [0,k],\ x \in N. \tag{9.3.2}$$

We leave the proof of the above lemma for the end of this section and we continue with the proof of (9.3.1).

It follows from (9.3.2) that

$$\int p_k(x,y)f(y)dy \le C\exp\left(-\frac{\left(\rho_k(x)+B\sqrt{k}\right)^2}{2Ak}\right) \tag{9.3.3}$$

for all $x \in N$ and $k \ge 1$.

On the other hand, it follows from (1.6.1) that there is $\beta > 1$ and $\lambda > 0$ such that

$$p_t(x,e) \le \lambda \ \inf\{p_{\beta t}(x,y), |y|_N \le \sqrt{t}\} \tag{9.3.4}$$

for all $x \in N$ and $t \ge 1$.
Since $|U^n| \le cn^D$, $n \in \mathbb{N}$ for some constant $c > 0$, it follows from (9.3.3) and (9.3.4) that

$$\begin{aligned} p_t(x,e) \le& \lambda \frac{1}{|U^{[\sqrt{t}]}|}\int_{U^{[\sqrt{t}]}} p_{\beta t}(x,y)dy \\ \le& \lambda c t^{-\frac{D}{2}} C\exp\left(-\frac{\left(\rho_{\beta t}(x)+B\sqrt{\beta t}\right)^2}{2A\beta t}\right) \end{aligned}$$

for all $x \in N$ and $t \ge 1$. This proves (10.3.1) and theorem 10.1 follows.

Proof of lemma 9.3.1. Let

$$m = \inf\{F(t,x):\ t \in [0,k],\ x \in N\}.$$

If $m \ge 0$, then there is nothing to prove. So, let us assume that $m < 0$.

Since for every neighborhood U of e,

$$\int_U p_t(e,y)dy \to 1 \ \text{ as } \ t \to 0 \tag{9.3.5}$$

we have that

$$\|u(t,.) - f\|_\infty \to 0 \ \text{ as } \ t \to 0.$$

So there is $\varepsilon \in (0,k)$ such that

$$F(t,x) \ge \frac{m}{2}, \quad (t,x) \in (0,\varepsilon) \times N.$$

Also, the semigroup property $p_{t+s} = p_t p_s$ and the left invariance $p_t(x,y) = p_t(e,x^{-1}y)$ of the heat kernel combined with (9.3.5) imply that there is $M > 0$ such that

$$u(t,x) = \int p_t(x,y)f(y)dy < \frac{|m|}{2}, \quad (t,x) \in [0,k] \times \{|x|_N > M\}$$

and hence

$$F(t,x) > \frac{m}{2}, \quad (t,x) \in [0,k] \times \{|x|_N > M\}.$$

It follows that

$$\inf\{F(t,x):\ \ (t,x) \in [0,k] \times N\} = \inf\{F(t,x):\ \ (t,x) \in [\varepsilon,k] \times \{|x|_N \le M\}\}.$$

Since $[\varepsilon,k] \times \{|x|_N \le M\}$ is compact, there is a point $(t_0,x_0) \in [\varepsilon,k] \times \{|x|_N \le M\}$ such that

$$F(t_0,x_0) = \inf\{F(t,x):\ \ (t,x) \in [0,k] \times N\} = m.$$

Since $h(x_0) = F(t_0, x_0)$ is the minimum of the function $h(x) = F(t_0, x)$, we have that

$$XF(t_0, x_0) = 0 \tag{9.3.6}$$
$$X^2F(t_0, x_0) \geq 0$$

for every vector field $X \in \mathfrak{n}$.
Since $L = -\left(E_1^2 + ... + E_p^2\right) + E_0$, it follows that

$$LF(t_0, x_0) \leq 0. \tag{9.3.7}$$

Also, since $F(t, x_0) \geq F(t_0, x_0)$, for $0 < t < t_0$, we have that

$$\frac{\partial}{\partial t} F(t_0, x_0) \leq 0. \tag{9.3.6}$$

It follows from (9.3.6) and (9.3.7) that

$$\left(\frac{\partial}{\partial t} + L_0\right) F(t_0, x_0) \leq 0.$$

This cannot hapen because by construction

$$\left(\frac{\partial}{\partial t} + L\right) F(t_0, x_0) = \left(\frac{\partial}{\partial t} + L\right) H_k(t_0, x_0) > 0.$$

This completes the proof of the lemma.

10. Polynomials on nilpotent Lie groups

The goal of this section is to prove proposition 1.9.1. We shall use the notation of sections 5 and 6.

Since we have identify N, as a differential manifold, with $\mathbb{R}^q$, using the exponential coordinates of the second kind, a monomial $P(t, x)$ on $\mathbb{R} \times N$ is just a monomial

$$P(t, x) = t^{i_0} x_1^{i_1} ... x_q^{i_q},$$

on $\mathbb{R} \times \mathbb{R}^q$.

Note that an equivalent definition of the homogeneous degree $\deg_H P$ of P is

$$\deg_H P = 2i_0 + i_1\sigma(1) + ... + i_q\sigma(q).$$

Let L be a centered left invariant sub-Laplacian on N and let L_0 be the associated limit sub-Lplacian. Then by (5.3.1) and (5.3.2)

$$\deg_H \left(\frac{\partial}{\partial t} + L\right) P(x) \leq \deg_H P(t, x) - 2 \tag{10.1}$$

and

$$\left(\frac{\partial}{\partial t} + L\right) P(t, x) = \left(\frac{\partial}{\partial t} + L_0\right) P(t, x) + Q(t, x), \tag{10.2}$$

where $Q(t, x)$ is a polynomial satisfying $\deg_H Q \leq \deg_H P - 3$.

We can also prove by induction on the dimension q of the Lie algebra $\mathfrak{n}$ of N and the homogeneous degree $\deg_H P$ of P that there is a polynomial $Q(t,x)$ satisfying

$$\left(\frac{\partial}{\partial t}+L_0\right)Q(t,x)=P(t,x) \tag{10.3}$$
$$\deg_H Q(t,x)=\deg_H P(t,x)+2.$$

Proposition 1.9.1 follows from (10.1), (10.2) and (10.3).

Note that if the polynomials Q_{P_i} are as in section 1.9.1, then for $0\le i\le \nu_2$, we can take $Q_{P_i}=P_i$. Note also that $\nu_0=0$ and that $\nu_1=n_1$. So, we shall assume that $P_0=Q_{P_0}=1$, that $P_i(t,x)=Q_{P_i}(t,x)=x_i$, for $1\le i\le \nu_1$ and that $P_i=Q_{P_i}$, for $\nu_1<i\le\nu_2$.

We shall also denote by $Q^{\varepsilon}_{P_i}$, $i\ge 0$ the monomials defined by

$$Q^{\varepsilon}_{P_i}(t,x)=\varepsilon^{\deg_H(P_i)}Q_{P_i}(\varepsilon^{-2}t,\tau_{1/\varepsilon}x),\quad \varepsilon>0$$

Note that

$$\left(\frac{\partial}{\partial t}+L_0\right)P_i(t,x)=\left(\frac{\partial}{\partial t}+L_\varepsilon\right)Q^{\varepsilon}_{P_i}(t,x) \tag{10.4}$$

and that

$$Q^{\varepsilon}_{P_i}(t,x)\to P_i(t,x),\quad \text{as } \varepsilon\to 0 \tag{10.5}$$

uniformly on the compact subsets of $\mathbb{R}\times\mathbb{R}^q$.

We shall say that a polynomial is homogeneous of degree k if it is a linear combination of monomials of homogeneous degree k. Note that if $W(t,x)$ is such a polynomial , then

$$W(\varepsilon^2 t,\tau_\varepsilon x)=\varepsilon^k W(t,x),\quad \varepsilon>0.$$

11. A Taylor formula for the heat functions on nilpotent Lie groups

The goal of this section is to prove theorem 1.9.2. The proof is based on the Harnack inequality (1.6.1) which provides a certain compactness on the space of positive heat functions and on a rescaling argument from Avellaneda and Lin [**AL1, AL2**].

11.1. The compactnes. Let us observe, with the notation of section 6, that if a function $u_\varepsilon,\varepsilon>0$ satisfies

$$\left(\frac{\partial}{\partial t}+L_\varepsilon\right)u_\varepsilon=0\quad\text{in}\quad(-1,1)\times B^0_1(e)$$

then the function

$$v_\varepsilon(t,x)=u_\varepsilon(\varepsilon^2 t,\tau_\varepsilon x)$$

satisfies

$$\left(\frac{\partial}{\partial t}+L\right)v_\varepsilon=0\quad\text{in}\quad(-\varepsilon^{-2},\varepsilon^{-2})\times B^0_{1/\varepsilon}(e).$$

So if $u_\varepsilon,\ 0<\varepsilon\le 1$ is a family of functions satisfying

$$\left(\frac{\partial}{\partial t}+L_\varepsilon\right)u_\varepsilon=0,\quad\text{in }(-1,1)\times B^0_1(e)$$
$$\|u_\varepsilon\|_\infty\le 1$$

then, by (1.6.5) and (6.2.1), there is a subsequence also denoted by u_ε and a function u_0, such that

$$u_\varepsilon \longrightarrow u_0, \quad (\varepsilon \to 0)$$

uniformly on the compact subsets of $(-1,1) \times B^0_1(e)$.

LEMMA 11.1.1. *Let u_0 and u_ε, $\varepsilon \in (0,1)$ be as above. Then*

$$\left(\frac{\partial}{\partial t} + L_0\right) u_0 = 0, \quad \text{in } (-1,1) \times B^0_1(e). \tag{11.1.1}$$

PROOF. We shall use the notation

$$\langle f, \phi \rangle = \int\int f(t,x)\phi(t,x)dxdt.$$

Let us consider a smooth function ϕ supported in $(-1,1) \times B^0_1(e)$. Then to prove (11.1.1) it is enough to prove that

$$\left\langle u_0, \left(-\frac{\partial}{\partial t} + L_0^*\right)\phi \right\rangle = 0 \tag{11.1.2}$$

where L^* is the adjoint of L.
We observe that by (6.1.1)

$$\left(-\frac{\partial}{\partial t} + L_\varepsilon^*\right)\phi = \left(-\frac{\partial}{\partial t} + L_0^*\right)\phi + \varepsilon\phi_\varepsilon$$

with ϕ_ε satisfying

$$\|\phi_\varepsilon\|_\infty \le c\|\nabla^2\phi\|_\infty, \quad 0 \le \varepsilon \le 1.$$

Therefore we have

$$\begin{aligned}
\left\langle u_0, \left(-\frac{\partial}{\partial t} + L_0^*\right)\phi \right\rangle &= \lim_{\varepsilon\to 0}\left\langle u_\varepsilon, \left(-\frac{\partial}{\partial t} + L_0^*\right)\phi \right\rangle \\
&= \lim_{\varepsilon\to 0}\left[\left\langle u_\varepsilon, \left(-\frac{\partial}{\partial t} + L_\varepsilon^*\right)\phi \right\rangle - \varepsilon\langle u_\varepsilon, \phi_\varepsilon\rangle\right] \\
&= \lim_{\varepsilon\to 0}\left[\left\langle \left(\frac{\partial}{\partial t} + L_\varepsilon\right)u_\varepsilon, \phi \right\rangle - \varepsilon\langle u_\varepsilon, \phi_\varepsilon\rangle\right] \\
&= -\lim_{\varepsilon\to 0}\varepsilon\langle u_\varepsilon, \phi_\varepsilon\rangle \\
&=0
\end{aligned}$$

which proves (11.1.2) and the lemma follows.

11.2. The Avellaneda-Lin rescaling argument. Let us denote by S_r the ball

$$S_r = (-r^2, r^2) \times \{x = (x_1, ..., x_q) : |x_i| \le r^i, 1 \le i \le q\}.$$

LEMMA 11.2.1. *For all $n \in \mathbb{N}$ there is a constant $c_n > 0$ such that for all functions u_0 satisfying*

$$\left(\frac{\partial}{\partial t} + L_0\right) u_0 = 0, \quad \text{in } S_1$$

we have

$$\sup\left\{|u_0 - \sum_{0\le i\le \nu_n} A_i^0 P_i|; S_\theta\right\} < c_n\theta^{n+1}\|u\|_\infty \tag{11.2.1}$$

for all $\theta \in (0,1)$ *and where the constants* A_i^0 *satisfy*

$$|A_i^0| \le c_n\|u\|_\infty \tag{11.2.2}$$

for all $1 \le i \le \nu_n$ *and*

$$\left(\frac{\partial}{\partial t} + L_0\right)\left(\sum_{\nu_{k-1}<i\le\nu_k} A_i^0 P_i\right) = 0 \tag{11.2.3}$$

for all $1 \le k \le n$ *.*

PROOF. If

$$P_i(t,x) = t^{i_0}x_1^{i_1}...x_q^{i_q}$$

then let us take as A_i^0

$$A_i^0 = \frac{\partial^{i_0}}{\partial t^{i_0}}\frac{\partial^{i_1}}{\partial x_1^{i_1}}\cdots\frac{\partial^{i_q}}{\partial x_q^{i_q}}u(0,0). \tag{11.2.4}$$

Then (11.2.1) follows from the usual Taylor formula and (11.2.2) from the local Harnack inequality (2.2.1). To prove (11.2.3) it is enough to prove that the homogeneous polynomials

$$W_k = \sum_{\nu_{k-1}<i\le\nu_k} A_i^0 P_i$$

with the A_i^0 as in (11.2.4), satisfy

$$\left(\frac{\partial}{\partial t} + L_0\right)W_k = 0. \tag{11.2.5}$$

We shall prove this by induction. Clearly (11.2.5) is true for $k = 1$. So let us assume that (11.2.5) is true for $k \le n-1$. We must prove that it is also true for $k = n$.

By the Taylor formula

$$\sup\left\{|u_0 - \sum_{0\le i\le n} W_i|; S_\theta\right\} < c_n\theta^{n+1}\|u\|_\infty \tag{11.2.6}$$

for all $\theta \in (0,1)$.

Let us assume for simplicity that $\|u\|_\infty \le 1$ and consider the function

$$v = u_0 - \sum_{0\le i\le n-1} W_i.$$

Then by the induction hypothesis

$$\left(\frac{\partial}{\partial t} + L_0\right)v = 0.$$

Also

$$\sup\{|v - W_n|; S_\theta\} < c_n\theta^{n+1} \tag{11.2.7}$$

for all $\theta \in (0,1)$. Let

$$v_\theta(t,x) = v(\theta^2 t, \tau_\theta x).$$

Then

$$\left(\frac{\partial}{\partial t} + L_0\right) v_\theta = 0$$

and by (11.2.7) and the homogeneity of W_n we have

$$\sup\{|v_\theta - \theta^n W_n|; S_1\} < c_n\theta^{n+1}$$

for all $\theta \in (0,1)$. It follows that

$$\sup\left\{|\theta^{-n}v_\theta - W_n|; S_1\right\} < c_n\theta$$

for all $\theta \in (0,1)$. So, W_n is a uniform limit of the functions $\theta^{-n}v_\theta$ as $\theta \to 0$ and hence it satisfies (11.2.5).

Lemma 11.2.2. *For all $n \in \mathbb{N}$ and $\mu \in (0,1)$ there are constants $c_n > 0$ and $\theta, \varepsilon_0 \in (0,1)$ such that for all $\varepsilon \in (0,\varepsilon_0)$ and all functions u_ε satisfying*

$$\left(\frac{\partial}{\partial t} + L_\varepsilon\right) u_\varepsilon = 0, \quad \textit{in } S_1$$

we have

$$\sup\left\{|u_\varepsilon - \sum_{0\le i\le \nu_n} A_i^\varepsilon Q_{P_i}^\varepsilon|; S_\theta\right\} < \theta^{n+\mu}\|u\|_\infty \tag{11.2.8}$$

where the constants A_i^ε satisfy

$$|A_i^\varepsilon| \le c_n\|u\|_\infty$$

for all $0 \le i \le \nu_n$ and

$$\left(\frac{\partial}{\partial t} + L_\varepsilon\right)\left(\sum_{\nu_{k-1}<i\le\nu_n} A_i^\varepsilon Q_{P_i}^\varepsilon\right) = 0$$

for all $1 \le k \le n$

Proof. By the previous lemma, there is $\mu' \in (\mu,1)$, $\theta \in (0,1)$ and $c_n > 0$ such that for all functions u_0 satisfying

$$\left(\frac{\partial}{\partial t} + L_0\right) u_0 = 0, \quad \text{in } S_1$$

we have

$$\sup\left\{|u_0 - \sum_{0\le i\le\nu_n} A_i^0 P_i|; S_\theta\right\} < \theta^{n+\mu'}\|u_0\|_\infty \tag{11.2.9}$$

where the constants A_i^0 satisfy

$$|A_i^0| \le c_n\|u_0\|_\infty$$

for all $1 \leq i \leq \nu_n$ and

$$\left(\frac{\partial}{\partial t} + L_0\right)\left(\sum_{\nu_{k-1} < i \leq \nu_n} A_i^0 P_i\right) = 0.$$

for all $1 \leq k \leq n$.

Let us fix these values of c_n and θ. If the lemma was false then then there would be a sequence of functions u_{ε_m} not satisfying (11.2.8). By considering the sequence $u'_{\varepsilon_m} = u_{\varepsilon_m}/\|u_{\varepsilon_m}\|_\infty$, we can assume that $\|u_{\varepsilon_m}\|_\infty \leq 1$. We can also assume, by extracting a subsequence if necessary, that $u_{\varepsilon_m} \to u_0$, as $m \to \infty$, uniformly on the compact subsets of S_1. Then, of course, u_0 would satisfy (11.2.9).

Let us take $A_i^{\varepsilon_m} = A_i^0, 0 \leq i \leq \nu_n$. Then using the assumption that the functions u_{ε_m} do not satisfy (11.2.8) and passing to the limit we have that

$$\theta^{n+\mu} < \sup\left\{|u_0 - \sum_{0 \leq i \leq \nu_n} A_i^0 P_i|; S_\theta\right\} < \theta^{n+\mu'}$$

which is absurd. Hence the lemma.

LEMMA 11.2.3. *Let μ, θ and ε_0 be as in the previous lemma. Then there is a constant $c_n > 0$ such that for all $m \in \mathbb{N}$ and $\varepsilon \in (0,1)$ such that $\varepsilon \leq \theta^{m-1}\varepsilon_0$ and all functions u_ε satisfying*

$$\left(\frac{\partial}{\partial t} + L_\varepsilon\right) u_\varepsilon = 0, \quad in\ S_1$$

we have

$$\sup\left\{|u_\varepsilon - \sum_{0 \leq i \leq \nu_n} A_i^{\varepsilon,m} Q_{P_i}^\varepsilon|; S_{\theta^m}\right\} < \theta^{m(n+\mu)}\|u\|_\infty \tag{11.2.10}$$

where the constants $A_i^{\varepsilon,m}$ satisfy

$$|A_i^{\varepsilon,m}| \leq c_n \|u\|_\infty$$

for all $0 \leq i \leq \nu_n$ and

$$\left(\frac{\partial}{\partial t} + L_\varepsilon\right)\left(\sum_{\nu_{k-1} < i \leq \nu_n} A_i^{\varepsilon,m} Q_{P_i}^\varepsilon\right) = 0$$

for all $1 \leq k \leq n$.

PROOF. We shall prove (11.2.10) by induction on m. For $m = 1$ we are in the case of lemma 11.2.2. So let us assume that (11.2.10) is true for some $m \in \mathbb{N}$. To prove that it is also true for $m+1$, let us assume for simplicity that $\|u_\varepsilon\|_\infty \leq 1$ and consider the function

$$w_\varepsilon(t,x) = \theta^{-m(n+\mu)}\left[u_\varepsilon(\theta^{2m}t, \tau_{\theta^m}x) - \sum_{0 \leq i \leq \nu_n} A_i^{\varepsilon,m} Q_{P_i}^\varepsilon(\theta^{2m}t, \tau_{\theta^m}x)\right].$$

Then

$$\left(\frac{\partial}{\partial t} + L_{\varepsilon\theta^{-m}}\right) w_\varepsilon = 0 \ \text{ in } \ S_1 \quad \text{and} \quad \|w_\varepsilon\|_\infty \le 1.$$

Since $\varepsilon\theta^{-m} \le \varepsilon_0$ it follows from lemma 11.2.2 that

$$\sup\left\{ |w_\varepsilon - \sum_{0\le i\le \nu_n} B_i^\varepsilon Q_{P_i}^{\varepsilon\theta^{-m}}|; S_\theta \right\} < \theta^{n+\mu}$$

where the constants B_i^ε satisfy

$$|B_i^\varepsilon| \le c_n$$

(with the constant c_n as in lemma 11.2.2) for all $0 \le i \le \nu_n$ and

$$\left(\frac{\partial}{\partial t} + L_{\varepsilon\theta^{-m}}\right)\left(\sum_{\nu_{k-1}<i\le \nu_n} B_i^\varepsilon Q_{P_i}^{\varepsilon\theta^{-m}} \right) = 0$$

for all $1 \le k \le n$. Let us observe that

$$\begin{aligned} Q_{P_i}^{\varepsilon\theta^{-m}}(t,x) &= \left(\varepsilon\theta^{-m}\right)^{\deg_H P_i} Q_{P_i}(\varepsilon^{-2}\theta^{2m}t, \tau_{\varepsilon^{-1}\theta^m}x) \\ &= \theta^{-m\deg_H P_i} Q_{P_i}^\varepsilon(\theta^{2m}t, \tau\theta^m x). \end{aligned}$$

So, if we set

$$A_i^{\varepsilon,m+1} = A_i^{\varepsilon,m} + \theta^{m(n+\mu-\deg_H P_i)} B_i^\varepsilon$$

then we have

$$\theta^{-m(n+\mu)} \sup\left\{ |u_\varepsilon(\theta^{2m}t, \tau_{\theta^m}x) - \sum_{0\le i\le \nu_n} A_i^{\varepsilon,m+1} Q_{P_i}^\varepsilon(\theta^{2m}t, \tau_{\theta^m}x)| : (t,x) \in S_\theta \right\} < \theta^{n+\mu}$$

and hence

$$\sup\left\{ |u_\varepsilon - \sum_{0<i<\nu_n} A_i^{\varepsilon,m+1} Q_{P_i}^\varepsilon|; S_{\theta^{m+1}} \right\} < \theta^{(m+1)(n+\mu)}$$

which proves the inductive step and the lemma follows.

Let us observe that every $\varepsilon \in (0,\varepsilon_0]$ satisfies $\theta^m\varepsilon_0 < \varepsilon \le \theta^{m-1}\varepsilon_0$ for some $m \in \mathbb{N}$. So the previous lemma has the following version:

LEMMA 11.2.4. *For all $n \in \mathbb{N}$ there is $\varepsilon_0 \in (0,1)$ and $c_n > 0$ such that for all $\varepsilon \in (0,\varepsilon_0]$ and all functions u_ε satisfying*

$$\left(\frac{\partial}{\partial t} + L_\varepsilon\right) u_\varepsilon = 0, \quad \text{in } S_1$$

we have

$$\sup\left\{ |u_\varepsilon - \sum_{0\le i\le \nu_n} A_i^\varepsilon Q_{P_i}^\varepsilon|; S_\varepsilon \right\} < c_n \varepsilon^{n+1} \|u\|_\infty \tag{11.2.11}$$

where the constants A_i^ε satisfy

$$|A_i^\varepsilon| \leq c_n \|u\|_\infty,$$

for all $0 \leq i \leq \nu_n$ and

$$\left(\frac{\partial}{\partial t} + L_\varepsilon\right)\left(\sum_{\nu_{k-1}<i\leq\nu_n} A_i^\varepsilon Q_{P_i}^\varepsilon\right) = 0$$

for all $1 \leq k \leq n$.

Note that (11.2.11) is trivially true for $\varepsilon \in (\varepsilon_0, 1]$. So, the restriction $\varepsilon \in (0, \varepsilon_0)$ in the above lemma can be replaced by $\varepsilon \in (0, 1]$.

11.3. A Taylor formula for heat functions. Let us observe again that if a function $u_\varepsilon, \varepsilon > 0$ satisfies

$$\left(\frac{\partial}{\partial t} + L_\varepsilon\right) u_\varepsilon = 0 \quad \text{in} \quad (-1,1) \times B_1^0(e)$$

then the function $v_\varepsilon(t,x) = u_\varepsilon(\varepsilon^2 t, \tau_\varepsilon x)$ satisfies

$$\left(\frac{\partial}{\partial t} + L\right) v_\varepsilon = 0 \quad \text{in} \quad (-\varepsilon^{-2}, \varepsilon^{-2}) \times B_{1/\varepsilon}^0(e)$$

So the following result is an immediate consequence of lemma 11.3.4.

THEOREM 11.3.1. *For all $n \in \mathbb{N}$ there is $c_n > 0$ such that for all $\Theta \geq 1$, $r \geq 1$ and all functions u satisfying*

$$\left(\frac{\partial}{\partial t} + L\right) u = 0, \quad in\ (-\Theta^2 r^2, \Theta^2 r^2) \times B_{\Theta r}(e)$$

we have

(11.3.1)

$$\sup\left\{|u - \sum_{0\leq i\leq\nu_n} A_i (\Theta r)^{-\deg_H P_i} Q_{P_i}|; (r^2, r^2) \times B_r(e)\right\} < c_n \Theta^{-(n+1)} \|u\|_\infty$$

where the constants A_i satisfy

$$|A_i| \leq c_n \|u\|_\infty$$

for all $0 \leq i \leq \nu_n$ and

$$\left(\frac{\partial}{\partial t} + L\right)\left(\sum_{\nu_{k-1}<i\leq\nu_k} A_i Q_{P_i}\right) = 0$$

for all $1 < k \leq n$.

Note that theorem 1.9.2 is a particular case of the above result.

12. Harnack inequalities for the derivatives of the heat functions on nilpotent Lie groups

It follows from theorem 11.3.1 and the local Harnack inequality (2.2.1) that for all $k, i_1, ..., i_q \in \mathbb{N}$ there is a constant $c > 0$ such that for all $r \geq 1$ and every function u satisfying $\left(\frac{\partial}{\partial t} + L\right) u = 0$ in $\left(-r^2, r^2\right) \times B_r(e)$ we have

$$(12.1) \qquad |\frac{\partial^k}{\partial t^k} \frac{\partial^{i_1}}{\partial x_1^{i_1}} \cdots \frac{\partial^{i_q}}{\partial x_q^{i_q}} u(0,e)| \leq c r^{-2k - i_1\sigma(1) - \ldots - i_q\sigma(q)} \|u\|_\infty .$$

The above inequality, combined with (5.3.1) and (3.1), implies the following :

THEOREM 12.1. *For for all $a, b > 0$ and all $k, n \in \mathbb{N}$, there are $\alpha, \beta > 1, \alpha < \beta$, $c > 0$ and $c_{k,n} > 0$ such that for all $r \geq 1$ and all $u \geq 0$ satisfying*

$$\left(\frac{\partial}{\partial t} + L\right) u = 0 \quad in \quad \left(0, (\beta + b^2) r^2\right) \times B_{cr}(e)$$

we have

$$(12.2) \quad \sup \left\{ |\frac{\partial^k}{\partial t^k} X_{i_1} X_{i_2} ... X_{i_n} u| ; \left(\alpha r^2, (\alpha + a^2) r^1\right) \times B_{ar}(e) \right\} \leq$$
$$c_{k,n} r^{-2k - \sigma(i_1) - \ldots - \sigma(i_n)} \inf \left\{ u; \left(\beta r^2, (\beta + b^2) r^2\right) \times B_{br}(e) \right\}$$

where $1 \leq i_j \leq q$, $1 \leq j \leq n$.

Applying the above result to the heat kernel $p_t(x, y)$ of L we have the following:

COROLLARY 12.2. *For all $k, n \in \mathbb{N}$ there is $c > 0$ such that*

$$(12.3) \quad |\frac{\partial^k}{\partial t^k} X_{i_1} X_{i_2} ... X_{i_n} p_t(x,y)| \leq c t^{-(D + 2k + \sigma(i_1) + \ldots + \sigma(i_n))/2} \exp\left(-\frac{|x^{-1}y|_N}{ct}\right)$$

for all $t \geq 1$ and $x, y \in N$.

13. Harmonic functions of polynomial growth on nilpotent Lie groups

The goal of this section is to give the proof of theorem 1.10.1. Note that in view of (2.1.2), a function u grows polynomially if and only if if there is $c > 0$ and $n \in \mathbb{N}$ such that

$$(13.1) \qquad \sup\{|u|; B_r(e)\} \leq c r^n, \quad r \geq 1$$

PROOF OF THEOREM 1.10.1. Let u be an L-harmonic function satisfying (13.1). Then by theorem 11.3.1 there is $c > 0$ and $r_0 > 1$ such that for all $r \geq r_0$

$$(13.2) \quad \sup \left\{ |u - \sum_{0 \leq i \leq \nu_n} A_i^r r^{-\deg_H P_i} Q_{P_i}|; B_1(e) \right\} < c_n r^{-(n+1)} \|u\|_{L^\infty(B_r(e))}$$

where we have set $Q_{P_i}(x) = Q_{P_i}(0, x)$ and where the constants A_i^r are such that

$$L\left(\sum_{0 \leq i \leq \nu_n} A_i^r r^{-\deg_H P_i} Q_{P_i} \right) = 0.$$

We set $C_{r,i} = 0$, for $Q_{P_i} = 0$ and

$$C_{r,i} = A_i^r r^{-\deg_H P_i}$$

otherwise. Then (13.1) and (13.2) imply that that there is $c > 0$ such that

$$\sup\left\{|u - \sum_{0\le i\le \nu_n} C_{r,i}Q_{P_i}|; B_1(e)\right\} < cr^{-1} \tag{13.3}$$

with

$$L\left(\sum_{0\le i\le \nu_n} C_{r,i}Q_{P_i}\right) = 0.$$

Now, there are constants C_i such that

$$C_{r,i} \to C_i \quad \text{as } r \to \infty \tag{13.4}$$

for all $0 \le i \le \nu_n$. To see this, let us observe that the there are constants $B_{r,i}$ such that

$$\sum_{0\le i\le \nu_n} C_{r,i}Q_{P_i} = \sum_{0\le i\le \nu_n} B_{r,i}P_i.$$

Now if

$$P_i(x) = x_1^{i_1}...x_q^{i_q}$$

then by the local Harnack inequality (2.2.1) we have that there is $c > 0$ such that

$$|B_{r,i} - \frac{\partial^{i_1}}{\partial x_1^{i_1}}...\frac{\partial^{i_q}}{\partial x_q^{i_q}}u(0)| \le \frac{c}{r}, \quad r \ge r_0$$

and hence

$$B_{r,i} \to \frac{\partial^{i_1}}{\partial x_1^{i_1}}...\frac{\partial^{i_q}}{\partial x_q^{i_q}}u(0), \quad \text{as } r \to \infty$$

for all $1 \le i \le \nu_n$. This implies (13.4), because the constants $B_{r,i}$ are linear combinations of the constant $C_{r,i}$ (in particular, if $\deg_H P_i = n$, then $C_{r,i} = B_{r,i}$).

By letting $r \to \infty$, it follows from (13.3) and (13.4) that

$$u(x) = \sum_{0\le i\le \nu_n} C_i Q_{P_i}(x)$$

for all $x \in B_1(x)$ and hence, by [**Bo** corollary 4.1], for all $x \in N$.

14. Proof of the Berry-Esseen estimate in the case of nilpotent Lie groups

The goal of this section is to prove theorem 1.14.1. We shall use the notation of sections 5 and 6.

Let us fix a C^∞ function $f \ge 0$ supported in $B_1^0(e)$ and satisfying $\int f(x)dx = 1$. Let $h(t,x) = p_t(x,e) - p_t^0(x,e)$ and set

$$h_f(t,x) = \int [p_t(x,y) - p_t^0(x,y)]f(y)dy.$$

Then, by (12.3),

$$\begin{aligned}|h(t,x)-h_f(t,x)| \le& \int |p_t(x,e)-p_t(x,y)|f(y)dy + \int |p^0_t(x,e)-p^0_t(x,y)|f(y)dy \\ \le& c\left(\|\nabla_X p_t(x,y)\|_\infty + \|\nabla_{0X} p^0_t(x,y)\|_\infty\right) \\ \le& ct^{-(D+1)/2}\end{aligned}$$

for all $t \ge 1$. So, to prove (1.14.1), it is enough to prove that

$$\|h_f(t,.)\|_\infty \le ct^{-(D+1)/2}, \quad t \ge 1. \tag{14.1}$$

Let

$$f_\varepsilon(x) = \frac{1}{\varepsilon^D} f(\tau_{\varepsilon^{-1}}x)$$

and

$$h^\varepsilon_f(t,x) = \int [p^\varepsilon_t(x,y) - p^0_t(x,y)] f_\varepsilon(y)dy.$$

Then to prove (14.1), it is enough to prove that

$$\|h^\varepsilon_f(1,.)\|_\infty \le c\varepsilon, \quad 0 < \varepsilon \le 1. \tag{14.2}$$

Let us consider the function

$$v_\varepsilon(t,x) = (\frac{\partial}{\partial t} + L_\varepsilon)h^\varepsilon_f(t,x).$$

Then

$$v_\varepsilon(t,x) = \int (L_\varepsilon - L_0)p^0_t(x,y)f_\varepsilon(y)dy.$$

Also, since

$$h^\varepsilon_f(0,x) = 0$$

we have

$$h^\varepsilon_f(1,x) = \int_0^1 \int p^\varepsilon_{1-t}(x,z)v_\varepsilon(t,z)dzdt.$$

So by (12.3) and (5.4.1) we have

$$\begin{aligned}|h^\varepsilon_f(1,x)| =& \int_0^1 \int p^\varepsilon_{1\ t}(x,z)v_\varepsilon(t,z)dzdt \\ =& \int_0^1 \int\int p^\varepsilon_{1-t}(x,z)(L_\varepsilon - L_0)p^0_t(z,y)f_\varepsilon(y)dydzdt \\ \le& c\,\varepsilon \left\{ \int_0^{1/2} \int\int \nabla^2_{\varepsilon X} p^\varepsilon_{1-t}(x,z)(1+|z|)^N p^0_t(z,y)f_\varepsilon(y)dydzdt \right. \\ &\left. + \int_{1/2}^1 \int\int p^\varepsilon_{1-t}(x,z)(1+|z|)^N \nabla^2_{0X} p^0_t(z,y)f_\varepsilon(y)dydzdt \right\} \\ \le& c\,\varepsilon \left\{ \int_0^{1/2} \|\nabla^2_{\varepsilon X} p^\varepsilon_{1-t}(x,.)\|_\infty \sup_{y\in V_\varepsilon} \|(1+|z|)^N p^0_t(z,y)\|_1 \|f_\varepsilon\|_1 dt \right. \\ &\left. + \int_{1/2}^1 \|p^\varepsilon_{1-t}(x,.)\|_1 \sup_{y\in V_\varepsilon} \|(1+|z|)^N \nabla^2_{0X} p^0_t(z,y)\|_\infty \|f_\varepsilon\|_1 dt \right\} \\ \le& c\varepsilon.\end{aligned}$$

This proves (14.2) and the theorem follows.

15. The nil-shadow of a simply connected solvable Lie group

Let S be a simply connected solvable Lie group. The goal of this section is to define the nil-shadow S_N of S.

Let $\mathfrak{s}$ be the Lie algebra of S which we identify with the left invariant vector fields on S.

Let us denote by $\mathfrak{s}$ the Lie algebra of S which we identify with the left invariant vector fields on S.

If $X \in \mathfrak{s}$ then we shall denote by $\mathcal{S}(X)$ and $\mathcal{N}(X)$ respectively the semisimple and nilpotent parts of the derivation $\operatorname{ad} X(Y) = [X, Y], X, Y \in \mathfrak{g}$.

Note that S has polynomial volume growth if and only if all the eigenvalues of the $\operatorname{ad} X$, $X \in \mathfrak{s}$ (hence also of the $\mathcal{S}(X)$, $X \in \mathfrak{s}$) are purely imaginary, i.e. of the form $ia, a \in \mathbb{R}$ (cf. [**Gui**]).

Let N be the nil-radical of S. This is the largest normal analytic nilpotent subgroup of S. It is also closed and $[S, S] \subseteq N$. Since S is simply connected, N is also simply connected and $S/N \cong \mathbb{R}^k$.

We denote by $\mathfrak{n}$ the Lie algebra of N.

LEMMA 15.1. *Let* $k = \dim(\mathfrak{s}/\mathfrak{n})$. *Then there are vector fields* $Y_1, ..., Y_k \in \mathfrak{s}$ *such that*

1. $S(Y_i)Y_j = 0$, $1 \leq i, j \leq k$, *and*
2. *the images of* $Y_1, ..., Y_k$ *under the projection* $\mathfrak{s} \to \mathfrak{s}/\mathfrak{n}$ *form a basis of* $\mathfrak{s}/\mathfrak{n}$.

An elementary proof of the above lemma was given in [**A1**]. We give below a more direct proof using the notion of Cartan subalgebras.

PROOF. Let us consider a Cartan subalgebra $\mathfrak{c}$ of $\mathfrak{s}$. $\mathfrak{c}$ is unique up to an inner automorphism (cf. [**Bou2** p.31]). Also $(\mathfrak{c} + \mathfrak{n})/\mathfrak{n}$ is a Cartan subalgebra of $\mathfrak{s}/\mathfrak{n}$ (cf. [**Bou2** p.20, **Ho** p. 170]) and hence $(\mathfrak{c} + \mathfrak{n})/\mathfrak{n} = \mathfrak{s}/\mathfrak{n}$. So there are vector fields $Y_1, ..., Y_k \in \mathfrak{c}$ which satisfy (2). Since these vector fields are elements of the Cartan subalgebra $\mathfrak{c}$, they automatically satisfy (1).

We fix vector fields $Y_1, ..., Y_k \in \mathfrak{s}$ as in the above lemma. Using these vector fields we can define on the linear space $\mathfrak{s}$ another Lie bracket $[.,.]_N$ by setting

$$[Y_i, Y_j]_N = [Y_i, Y_j], \;\; [Y_i, Y]_N = \mathcal{N}(Y_i)Y, 1 \leq i, j \leq k, \;\; Y \in \mathfrak{n}.$$

Making use of the fact that $\mathcal{S}(X)$ and $\mathcal{N}(X)$ are derivations, we can easily prove that $[.,.]_N$ satisfies the Jacobi identity and therefore $\mathfrak{s}_N = (\mathfrak{s}, [.,.]_N)$ is a nilpotent Lie algebra.

DEFINITION. We call $\mathfrak{s}_N$ the nil-shadow of $\mathfrak{s}$.

The map

$$\mathfrak{n} \times \mathbb{R}^k \to S \;:\; (Y, t_k, ..., t_1) \to \exp Y \exp t_k Y_k ... \exp t_1 Y_1$$

is an analytic isomorphism. Using this map, we shall identify S as a differential manifold with $\mathfrak{n} \times \mathbb{R}^k$.

Now, we can define on S another product $*_{S_N}$. This is done by setting

$$g *_{S_N} h = gh, \quad \text{for } g, h \in N$$

$$g_i *_{S_N} g_j = g_i g_j,$$

for $g_i = \exp t_i Y_i,\ g_j = \exp t_j Y_j,\ t_i, t_j \in \mathbb{R}, 1 \leq i, j \leq k$ and

$$g_i *_{S_N} h *_{S_N} g_i^{-1} = \exp e^{t_i \mathcal{N}(Y_i)} Y,$$

for $h = \exp Y,\ Y \in \mathfrak{n},\ g_i = \exp t_i Y_i,\ t_i \in \mathbb{R}, 1 \leq i \leq k.$

Note that the group $S_N = (S, *_N)$ is nilpotent and that its Lie algebra is isomorphic to $\mathfrak{s}_N$.

DEFINITION. We call S_N the nil-shadow of S.

If $g = \exp Y \exp t_k Y_k ... \exp t_1 Y_1$, then we shall denote by ϑ_g the Lie algebra automorphism of $\mathfrak{s}$ defined by

$$\vartheta_g = e^{t_k \mathcal{S}(Y_k)} ... e^{t_1 \mathcal{S}(Y_1)}.$$

Note that the automorphisms $\vartheta_g, g \in S$ are semisimple and commute. Note also that they are Lie algebra automorphisms of $\mathfrak{s}_N$.

Let us denote by $*_S$ the product of S (to distinguish it from the product $*_{S_N}$ of S_N). These products are related as follows: Let $g = \exp X \exp(t_k Y_k) ... \exp(t_1 Y_1)$ and $h = \exp Y \exp(s_k Y_k) ... \exp(t_1 Y_1)$ with $X, Y \in \mathfrak{n}$. Then, by the above definitions

$$\text{(15.1.)} \quad g *_S h = (\exp X \exp t_k Y_k ... \exp t_1 Y_1) *_{S_N} (\exp \vartheta_g Y \exp s_k Y_k ... \exp t_1 Y_1).$$

The following lemma is an immediate consequence of (15.1).

LEMMA 15.2. *Let $Y \in \mathfrak{s}$ and $X \in \mathfrak{s}_N$ be $*_S$ and $*_{S_N}$ left invariant vector fields respectively, satisfying $Y(e) = X(e)$. Then*

$$\text{(15.2)} \qquad Y(g) = (\vartheta_g X)(g)$$

for all $g \in S_N$.

Note that by (15.2) the vector fields $Y_1, ..., Y_k$ are $*_{S_N}$ as well as $*_S$-left invariant.

Note also that (1.4.1) is a consequence of (15.2).

15.1. Comparison of $|.|_S$ and $|.|_{S_N}$. Let us assume that S has polynomial volume growth, let W be a compact and convex neighbourhood of the origin in $\mathfrak{n}$, let

$$V = \cup_{g \in S} \vartheta_g W$$

and let

$$U = \{\exp Y \exp t_k Y_k ... \exp t_1 Y_1 : Y \in \overline{V}, |t_i| \leq 1, 1 \leq i \leq k\}.$$

Then U is a compact neighbourhood of the identity element e of S. Furthermore, by (16.1)

$$\text{(15.1.1)} \qquad g *_S U = g *_{S_N} U, \quad g \in S.$$

Let

$$U^{*_S n} = \{g_1 *_S g_2 *_S ... *_S *_S g_n : g_1, g_2, ..., g_n \in U\}$$
$$U^{*_{S_N} n} = \{g_1 *_{S_N} g_2 *_{S_N} ... *_{S_N} g_n : g_1, g_2, ..., g_n \in U\}.$$

Then, by (15.1.1)

$$\text{(15.1.2)} \qquad U^{*_S n} = U^{*_{S_N} n}.$$

Note that this shows that the groups S and S_N have the same homogeneous dimension D.

Let us define $|g|_S$ and $|g|_{S_N}$ as in (1.1), i.e. by setting

$$|g|_S = \inf\{n : g \in U^{*_S n}\}$$
$$|g|_{S_N} = \inf\{n : g \in U^{*_{S_N} n}\}.$$

Then by (16.1.2)

$$|g|_S = |g|_{S_N}, \qquad g \in S. \tag{15.1.3}$$

LEMMA 15.1.1. *For all $g, h \in S$*

$$|g^{-1*_S} *_S h|_S = |g^{-1*_{S_N}} *_{S_N} h|_{S_N} \tag{16.1.4}$$

where by g^{-1*_S} and $g^{-1*_{S_N}}$ we denote respectively the inverses of g with respect to the group products $*_S$ and $*_{S_N}$.

PROOF. We have that

1. $g^{-1*_S} *_S \in U^{*_S n}$ if and only if $h \in g *_S U^{*_S n}$ and
2. $g^{-1*_{S_N}} *_{S_N} h \in U^{*_{S_N} n}$ if and only if $h \in g *_{S_N} U^{*_{S_N} n}$.

On the other hand, by (15.1.1) and (15.1.2)

$$g *_S U^{*_S n} = g *_{S_N} U^{*_{S_N} n}, \qquad n \in \mathbb{N}.$$

Hence the lemma.

COROLLARY 15.1.2. *Let $d_L(.,.)$ be the control distance associated to a $*_S$-left invariant sub-Laplacian L on S. Then, there is $c > 0$ such that for all $g, h \in S$*

$$\frac{1}{c}|g^{-1*_{S_N}} h|_{S_N} \leq 1 + d_L(g, h) \leq c|g^{-1*_{S_N}} h|_{S_N}. \tag{15.1.5}$$

16. Connected Lie groups of polynomial volume growth

Let G be a connected Lie group of polynomial volume growth and let us fix a centered left inavariant sub-Laplacian $L = -(E_1^2 + ... + E_p^2) + E_0$ on G. The goal of this section is to define the homogenised operator L_H associated to L. The general strategy is the same as in section 1.5.

This section is rather technical. The reader who wants to avoid all these algebraic technicalities can just assume that $G = \mathbb{R}^k \rtimes M$ (this case is explained in section 1.5) and skip this section.

We shall adopt the notation of section 1.4.7. Since we are going to consider the projection of L on G/C, we shall assume that C is trivial. Then we can also assume that $G = S_1 \rtimes M_1$. Note also that $G/N \cong (S_1/N) \times M_1 \cong \mathbb{R}^k \times M_1$.

Let us denote by $\mathfrak{g}$ the Lie algebra of G which we identify with the left invariant vector fields on G. Let us also denote by $(\mathfrak{s})_1$, $\mathfrak{n}$ and $\mathfrak{m}_1$ the Lie algebras of S_1, N and M_1 respectively.

As in the previous section, if $X \in \mathfrak{g}$, then we shall denote by $\mathcal{S}(X)$ and $\mathcal{N}(X)$ respectively the semisimple and nilpotent parts of the derivation $\operatorname{ad} X(Y) = [X, Y], X, Y \in \mathfrak{g}$. The assumption that G has polynomial volume growth implies that all the eigenvalues of $\operatorname{ad} X$, $X \in \mathfrak{g}$ (hence also of $\mathcal{S}(X)$, $X \in \mathfrak{g}$) are purely imaginary, i.e. of the form $ia, a \in \mathbb{R}$ (cf. [**Gui**]).

LEMMA 16.1 (cf. [**A2**]). *Let* $k = \dim(\mathfrak{s}_1/\mathfrak{n})$. *Then there are* $Y_1, ..., Y_k \in \mathfrak{s}$ *such that*

1. $S(Y_i)Y_j = 0, \ 1 \leq i, j \leq k,$
2. $[Z, Y_i] = 0, \ 1 \leq i \leq k, \ Z \in \mathfrak{m}_1$ *and*
3. *the images of* $Y_1, ..., Y_k$ *under the projection* $\mathfrak{s}_1 \to \mathfrak{s}_1/\mathfrak{n}$ *form a basis of* $\mathfrak{s}/\mathfrak{n}$.

The above lemma can also be proved in the same way as lemma 15.1. The only change is that the algebra $\mathfrak{c}$ considered, must be a Cartan subalgebra of the centraliser of $\mathfrak{m}_1$ in $\mathfrak{s}_1$.

We fix vector fields $Y_1, ..., Y_k \in \mathfrak{s}_1$ as in the above lemma.

Using these vector fields, we define in exactly the same way as in section 15 the nil-shadow $(\mathfrak{s}_1)_N = (\mathfrak{s}_1, [.,.]_N)$ of $\mathfrak{s}_1$.

Using the analytic isomorphism

$$\mathfrak{n} \times \mathbb{R}^k \to S \ : \ (Y, t_k, ..., t_1) \to \exp Y \exp t_k Y_k ... \exp t_1 Y_1$$

we identify S_1 as a differential manifold with $\mathfrak{n} \times \mathbb{R}^k$.

Finally, we define in the same way the nil-shadow $(S_1)_N$ of S_1, which is also identified as a differential manifold with $\mathfrak{n} \times \mathbb{R}^k$.

If $g = (\exp Y \exp t_k Y_k ... \exp t_1 Y_1, z)$, then we shall denote by ϑ_g the Lie algebra automorphism of $\mathfrak{s}_1$ defined by

$$\vartheta_g = e^{t_k S(Y_k)} ... e^{t_1 S(Y_1)} \operatorname{Ad} z.$$

Note that ϑ_g is also a Lie algebra automorphism of $(\mathfrak{s}_1)_N$.

We shall extend the automorphisms $\vartheta_g, g \in G$ to $\mathfrak{g}$ by setting $\vartheta_g Z = Z, \ Z \in \mathfrak{m}_1$. Note that these automorphisms are semisimple.

Let us denote by G_N the group $G_N = (S_1)_N \times M_1$ and by $\mathfrak{g}_N$ its Lie algebra. Note that both of the groups G_N and G are identified as differential manifolds with $\mathfrak{n} \times \mathbb{R}^k \times M_1$.

Let us denote by $*_{G_N}$ the group product of G_N. Let us also denote by $*_{(S_1)_N}$ the group product of $(S_1)_N$ and by $*_G$ the group product of G (to distinguish it from the product $*_{G_N}$).

These products are related as follows: Let

$$g = (\exp X \exp t_k Y_k ... \exp t_1 Y_1, z), \qquad h = (\exp Y \exp s_k Y_k ... \exp t_1 Y_1, w) \in G.$$

Then,

$$\begin{aligned}(16.1) \quad & g *_G h = \\ & \big((\exp X \exp t_k Y_k ... \exp t_1 Y_1) *_{(S_1)_N} (\exp \vartheta_g Y \exp s_k Y_k ... \exp t_1 Y_1) \, , zw \big).\end{aligned}$$

The following lemma is an immediate consequence of (16.1).

LEMMA 16.2. *Let* $Y \in \mathfrak{g}$ *and* $X \in \mathfrak{g}_N$ *be* $*_G$ *and* $*_{G_N}$*-left invariant vector fields respectively, satisfying* $Y(e) = X(e)$. *Then*

$$Y(g) = (\vartheta_g X)(g) \tag{16.2}$$

for all $g \in G_N$.

Note that by (16.2) the vector fields $Y_1, ..., Y_k$ are both $*_{G_N}$ and $*_G$-left invariant.

16.1. Comparison of the controll distances on G and $(S_1)_N$. Let A be a compact and convex neighbourhood of the origin in $\mathfrak{n}$, let

$$W = \cup_{g\in G}\vartheta_g A$$

and set

$$V = \{\exp Y \exp t_k Y_k ... \exp t_1 Y_1 : Y \in \overline{W}, |t_i| \leq 1, 1 \leq i \leq k\}$$

and

$$U = V \times M_1.$$

Then U is a compact neighbourhood of the identity element e of G. Furthermore, by (16.1),

$$g *_G U = g *_{G_N} U, \quad g \in S. \tag{16.1.1}$$

Let

$$U^{*_G n} = \{g_1 *_G g_2 *_G ... *_G *_G g_n : g_1, g_2, ..., g_n \in U\}$$
$$U^{*_{G_N} n} = \{g_1 *_{G_N} g_2 *_{G_N} ... *_{G_N} g_n : g_1, g_2, ..., g_n \in U\}$$
$$V^{*_{(S_1)_N} n} = \{x_1 *_{(S_1)_N} x_2 *_{(S_1)_N} ... *_{(S_1)_N} x_n : x_1, x_2, ..., x_n \in V\}.$$

Then, by (16.1) and (16.1.1)

$$U^{*_G n} = U^{*_{G_N} n} = V^{*_{(S_1)_N} n} \times M_1. \tag{16.1.2}$$

This shows, in particular, that the groups G and $(S_1)_N$ have the same homogeneous dimension D.

Let us define $|g|_G$, $|g|_{G_N}$ and $|x|_{(S_1)_N}$ as in (1.1), i.e. by setting

$$|g|_G = \inf\{n : g \in U^{*_G n}\}$$
$$|g|_{G_N} = \inf\{n : g \in U^{*_{G_N} n}\}$$
$$|x|_{(S_1)_N} = \inf\{n : x \in V^{*_{(S_1)_N} n}\}.$$

If $g = (x, z)$, then by (16.1.2)

$$|x|_{(S_1)_N} = |g|_G = |g|_{G_N}. \tag{16.1.3}$$

LEMMA 16.1.1. *For all $g, h \in G$*

$$|g^{-1*_G} *_G h|_G = |g^{-1*_{G_N}} *_{G_N} h|_{G_N} \tag{16.1.4}$$

where by g^{-1*_G} and $g^{-1*_{G_N}}$ we denote the inverses of g with respect to the group products $*_G$ and $*_{G_N}$ respectively.

The above lemma can be proved in the same as lemma 15.1.1.

COROLLARY 16.1.2. *Let $d_L(.,.)$ be the control distance associated to a $*_G$-left invariant sub-Laplacian L on G. Then, there is $c > 0$ such that for all $g = (x, z), h = (y, w) \in G$*

$$\frac{1}{c}|x^{-1*_{(S_1)_N}} y|_{(S_1)_N} \leq 1 + d_L(g, h) \leq c|x^{-1*_{(S_1)_N}} y|_{(S_1)_N}. \tag{16.1.5}$$

16.2. The filtration and the basis of $(\mathfrak{s}_1)_N$. Since $(\mathfrak{s}_1)_N$ is nilpotent, we can proceed as in section 6.1 and obtain the filtration

$$(\mathfrak{s}_1)_N = (\mathfrak{s}_1)_{N1} \supseteq (\mathfrak{s}_1)_{N2} \supseteq ... \supseteq (\mathfrak{s}_1)_{Nm} \supseteq (\mathfrak{s}_1)_{N(m+1)} = \{0\}, \ (\mathfrak{s}_1)_{Nm} \neq \{0\}.$$

by setting $(\mathfrak{s}_1)_{N1} = (\mathfrak{s}_1)_N$ and $(\mathfrak{s}_1)_{N(i+1)} = [(\mathfrak{s}_1)_{N1}, (\mathfrak{s}_1)_{Ni}]_N$, $i \geq 1$. Note that $(\mathfrak{s}_1)_{N2} \subseteq \mathfrak{n}$.

Since the $\mathcal{S}(Y_i)$, $1 \leq i \leq k$, are semisimple and commute between them and with the $\operatorname{ad} Z$, $Z \in \mathfrak{m}_1$, we can consider linear subspaces $\mathfrak{a}_1, ..., \mathfrak{a}_m$ of $(\mathfrak{s}_1)_N$ such that

1. $(\mathfrak{s}_1)_{Ni} = \mathfrak{a}_i \oplus ... \oplus \mathfrak{a}_m$, $\quad 1 \leq i \leq m$,
2. $\mathcal{S}(Y_j)\mathfrak{a}_i \subseteq \mathfrak{a}_i$, $1 \leq i \leq m$, $1 \leq j \leq k$,
3. $\operatorname{ad} Z(\mathfrak{a}_i) \subseteq \mathfrak{a}_i$, $Z \in \mathfrak{m}_1$,, and
4. $\mathfrak{a}_1$ is spanned by $\mathfrak{a}_1 \cap \mathfrak{n}$ and the vector fields $\{Y_1, ..., Y_k\}$.

We set

$$n_0 = 0, \ n_i = \dim(\mathfrak{a}_1 \oplus ... \oplus \mathfrak{a}_i), \ 1 \leq i \leq m$$
$$\sigma(j) = i, \ for \ n_{i-1} < j \leq n_i$$
$$q = n_m = \dim(\mathfrak{s}_1)_N = \dim(\mathfrak{s}_1).$$

Notice that the homogeneous dimension D of G, which by (16.1.2) is equal to the homogeneous dimension of $(S_1)_N$, is given by

$$D = \sigma(1) + ... + \sigma(q).$$

We consider a basis $\{X_1, ..., X_q\}$ of $(\mathfrak{s}_1)_N$ such that

1. $Y_i(e) = X_i(e)$ (and hence $Y_i(g) = X_i(g)$, $g \in G_N$), $1 \leq i \leq k$,
2. $\{X_{k+1}, ..., X_q\} \subseteq \mathfrak{n}$ and
3. $\{X_{n_{i-1}+1}, ..., X_{n_i}\}$ is a basis of $\mathfrak{a}_i$, $1 \leq i \leq m$.

Furthermore, since the eigenvalues of the $\mathcal{S}(Y_\ell)$ are of the type $ia, a \in \mathbb{R}$, we can assume that each X_i,

1. either satisfies

$$\mathcal{S}(Y_\ell)X_i = 0, \ \ 1 \leq \ell \leq k, \tag{16.2.1}$$

2. or it can be combined with either its previous or its next vector X_j (i.e. either $j = i-1$ or $j = i+1$) so that for every $1 \leq \ell \leq k$ and all $t \subset \mathbb{R}$

$$\begin{aligned} e^{t\mathcal{S}(Y_\ell)}X_i &= \cos a_\ell t \ X_i + \sin a_\ell t \ X_j \\ e^{t\mathcal{S}(Y_\ell)}X_j &= -\sin a_\ell t \ X_i + \cos a_\ell t \ X_j \end{aligned} \tag{16.2.2}$$

for some $a = (a_1, ..., a_k) \in \mathbb{R}^k$, $a \neq 0$.

Again, by the fact that the $\mathcal{S}(Y_i)$, $1 \leq i \leq k$, are semisimple and commute between them and with the $\operatorname{ad} Z$, $Z \in \mathfrak{m}_1$, we have that each subspace $\mathfrak{a}_i$ splits into a direct sum

$$\mathfrak{a}_i = \mathfrak{a}_i^0 \oplus \mathfrak{a}_i^1 \tag{16.2.3}$$

with

$$\mathfrak{a}_i^0 = \{X \in \mathfrak{a}_i : \mathcal{S}(Y_j)X = 0, 1 \leq j \leq k \text{ and } \operatorname{ad} Z(X) = 0, \ Z \in \mathfrak{m}_1\}$$

and

$$\mathcal{S}(Y_j)\mathfrak{a}_i^1 \subseteq \mathfrak{a}_i^1, \ 1 \leq j \leq k \quad \text{and} \quad \operatorname{ad} Z(\mathfrak{a}_i^1) \subseteq \mathfrak{a}_i^1, \ Z \in \mathfrak{m}_1.$$

Note that $X_1, ..., X_k \in \mathfrak{a}_1^0$ and that $\mathfrak{a}_1^1 \subseteq \mathfrak{n}$.

We shall assume that the vector fields $\{X_1, ..., X_q\}$ have been chosen in such a way that for each $1 \leq i \leq m$ the vector fields $\{X_{n_{i-1}+1}, ..., X_{k_i}\}$ form a basis of $\mathfrak{a}_i^0$ and the vector fields $\{X_{k_i+1}, ..., X_{n_i}\}$ a basis of $\mathfrak{a}_i^1$.

Note that $k \leq k_1$. Note also that the vector fields $\{X_{k+1}, ..., X_{k_1}\}$ are also $*_G$-left invariant.

We shall denote by $Y_1, ..., Y_q$ the $*_G$-left invariant vector fields satisfying $Y_i(e) = X_i(e)$, $1 \leq i \leq q$.

Note that if $1 \leq i \leq k_1$, then $Y_i(g) = X_i(g)$, $g \in G$.

Note also that $\{Y_{k_1+1}, ..., Y_q\}$ is a basis of $[\mathfrak{s}, \mathfrak{g}]$. Therefore, a left inavariant sub-Laplacian $L = -(E_1^2 + ... + E_p^2) + E_0$ on G is centered if and only if $E_0 \in \mathfrak{m}_1 \oplus \mathfrak{a}_1^1 \oplus (\mathfrak{s}_1)_{N2}$.

16.3. The matrix of ϑ_g. Let us denote by ϑ_{gij} the entries of the matrix of ϑ_g with respect to the basis $\{X_1, ..., X_q\}$.

By (16.2.1) and (16.2.2), they have the following properties:

1. They are functions of type QP.
2. If either $n_{\ell-1} < i \leq k_\ell$ or $n_{\ell-1} < j \leq k_\ell$ for some $1 \leq \ell \leq m$, then

$$\vartheta_{gij} = \delta_{ij} \tag{16.3.1}$$

where $\delta_{ij} = 1$ for $i = j$ and $\delta_{ij} = 0$ for $i \neq j$ and

3. if $k_\ell < i, j \leq n_\ell$, for some $1 \leq \ell \leq m$, then

$$\langle \vartheta_{gij} \rangle = 0. \tag{16.3.2}$$

16.4. The sub-Laplacian L written as an operator on G_N. Let us consider a basis $\{X_{-r}, ..., X_0\}$ of $\mathfrak{m}_1$. Then $\{X_{-r}, ..., X_0, X_1, ..., X_q\}$ is a basis of the Lie algebra $\mathfrak{g}_N$ of G_N.

Let $L = -(E_1^2 + ... + E_p^2) + E_0$ be a centered $*_G$-left invariant sub-Laplacian on G. Then, by (16.2), (16.3.1) and (16.3.2) L can be written as

$$L = -\sum_{-r \leq i,j \leq q} X_i a_{ij} X_j + \sum_{-r \leq i \leq q} a_i X_i + \sum_{k_1 < i \leq q} a_i X_i \tag{16.4.1}$$

where the coefficients a_{ij} and a_i are functions of type QP with the following properties:

$$\begin{aligned} a_{ij} &= a_{ji}, & -r \leq i, j \leq q \\ a_{ij} &= \text{const.}, & -r \leq i, j \leq k \\ a_i &= 0, & 1 \leq i \leq k_1 \\ \langle a_i \rangle &= 0, & k_\ell < i \leq n_\ell. \end{aligned} \tag{16.4.2}$$

16.5. The correctors and the homogenised operator L_H. Let us fix a function $f \in C^\infty((S_1)_N)$ and let us extend f to G_N by setting $f(x, z) = f(x), (x, z) \in G_N = (S_1)_N \times M_1$. Then, by (16.4.1), there is a constant $c > 0$ independent of f such that

$$Lf = -\sum_{1 \leq i,j \leq n_1} a_{ij} X_i X_j f - \sum_{\substack{-r \leq i \leq k \\ 1 \leq j \leq n_2}} (X_i a_{ij}) X_j f + \sum_{1 \leq i \leq n_2} a_i X_i f + F \tag{16.5.1}$$

with the function F satisfying $|F(x,z)| \leq c\nabla_X^3 f(x)$, $(x,z) \in G_N$.

DEFINITION. We define the (first order) correctors $\psi^j, 1 \leq j \leq n_1$ (cf. [**BLP, JKO**]) as functions type QP satisfying

$$L\psi^j = -a_j + \sum_{-r\leq i\leq k} X_i a_{ij}, \quad \langle \psi^j \rangle = 0. \tag{16.5.2}$$

Note that $\langle X_i a_{ij} \rangle = 0$ and that by (16.4.2), $\langle a_j \rangle = 0$, for $k_1 < j \leq n_1$. So, the correctors ψ^j, $j = 1, ..., n_1$ are well defined.

Note also that $\psi^j = 0$, for $1 \leq j \leq k_1$.

Combining (16.5.1) and (16.5.2) we have

$$\begin{aligned} &L\left(f + \sum_{1\leq j\leq n_1} \psi^j X_j f\right) \\ &= -\sum_{1\leq i,j\leq n_1}\left(a_{ij} + \sum_{-r\leq \ell\leq k} a_{i\ell}X_\ell\psi^j + \sum_{-r\leq \ell\leq k} X_\ell\left(a_{\ell i}\psi^j\right)\right) X_iX_jf \\ &\quad + \sum_{1\leq i,j\leq n_1} a_i\psi^j X_iX_jf - \sum_{\substack{-r\leq i\leq k \\ n_1<j\leq n_2}} \left(X_ia_{ij}\right)X_jf \\ &\quad + \sum_{n_1<i\leq n_2} a_iX_if + F \end{aligned} \tag{16.5.3}$$

with the function F satisfying $|F(x,z)| \leq c\nabla_X^3 f(x)$, $(x,z) \in G_N$.

The following definitions are motivated by (16.5.3).

DEFINITION. The homogenised operator L_H associated to L is defined to be the operator

$$L_H = -\sum_{1\leq i,j\leq n_1} q_{ij}X_iX_j + \sum_{n_1<i\leq k_2} q_iX_i$$

with coefficients defined by

$$q_i = \left\langle a_i - \sum_{-r\leq i\leq k} X_ia_{ij}\right\rangle, \qquad n_1 < i \leq n_2$$

$$q_{ij} = \left\langle a_{ij} - a_i\psi^j + \sum_{-r\leq \ell\leq k} a_{i\ell}X_\ell\psi^j\right\rangle, \qquad 1 \leq i,j \leq n_1.$$

Note that $q_i = a_i$, $n_1 < i \leq k_2$.

DEFINITION. We define the (second order) correctors $\psi^{ij}, 1 \leq i,j \leq n_1$ (cf. [**BLP, JKO**]) as functions of type QP satisfying

$$L\psi^{ij} = a_{ij} - a_i\psi^j + \sum_{-r\leq \ell\leq k} a_{i\ell}X_\ell\psi^j + \sum_{-r\leq \ell\leq k} X_\ell\left(a_{\ell i}\psi^j\right) - q_{ij}, \quad \langle \psi^{ij}\rangle = 0. \tag{16.5.4}$$

We also define the (second order) correctors $\psi^j, n_1 < j \leq n_2$ as functions of type QP satisfying

$$L\psi^j = -a_j + \sum_{-r\leq i\leq k} X_ia_{ij}, \quad \langle \psi^j \rangle = 0. \tag{16.5.5}$$

Note that $\psi^j = 0$, for $n_1 < j \leq k_2$.

The following lemma explains the relation betweem L and L_H. It is a direct consequence of (16.5.3) and the above definitions.

LEMMA 16.5.1. *Let $f \in C^\infty((S_1)_N)$ and let us extend f to G_N by setting $f(x,z) = f(x), (x,z) \in G_N = (S_1)_N \times M_1$. Then, there is a constant $c > 0$ independent of f such that*

$$L\left(f + \sum_{1\leq j\leq n_2} \psi^j X_j f + \sum_{1\leq i,j\leq n_1} \psi^{ij} X_j f\right) = L_H f + F \tag{16.5.6}$$

with the function F satisfying $|F(x,z)| \leq c\left(\nabla_X^3 f(x) + \nabla_X^4 f(x)\right), (x,z) \in G_N$.

16.6. L_H is a centered sub-Laplacian on $(S_1)_N$. The following lemma shows that L_H is indeed a hypoelliptic operator.

LEMMA 16.6.1. *For all $\xi = (\xi_1, ..., \xi_{n_1}) \in \mathbb{R}^{n_1}, \xi \neq 0$,*

$$\sum_{1\leq i,j\leq n_1} q_{ij}\xi_i\xi_j > 0. \tag{16.6.1}$$

PROOF. The proof is similar to the one given in section 1.4.3. Let us recall that $(S_1)_N$ has been identified to $\mathbb{R}^q$ and let us denote by P_i, $1 \leq i \leq n_1$ the monomials $P_i(x,z) = x_i$, $(x,z) \in G_N = (S_1)_N \times M_1$. Then it follows from (16.5.2) that

$$L\left(P_i + \psi^i\right) = 0, \quad 1 \leq i \leq n_1. \tag{16.6.2}$$

Let us fix a vector $\xi = (\xi_1, ..., \xi_{n_1}) \neq 0$ and consider the function

$$u = \sum_{1\leq i\leq n_1} \xi_i \left(P_i + \psi^i\right).$$

Then, by (16.6.2), $Lu = 0$. Furthermore

$$Lu^2 = 2uLu - 2[(E_1u)^2 + ... + (E_pu)^2] = -2[(E_1u)^2 + ... + (E_pu)^2] \leq 0. \tag{16.6.3}$$

This implies that Lu^2 is a function of type QP. We want to prove that

$$\langle Lu^2 \rangle = -2 \sum_{1\leq i,j\leq n_1} q_{ij}\xi^i\xi^j. \tag{16.6.4}$$

We have

$$\begin{aligned} u^2 &= \sum_{1\leq i,j\leq n_1} \xi_i\xi_j(P_i + \psi^i)(P_j + \psi^j) \\ &= \sum_{1\leq i,j\leq q} \xi_i\xi_j(P_iP_j + P_i\psi^j + P_j\psi^i + \psi^i\psi^j). \end{aligned}$$

Since

$$L(P_iP_j) = \left(a_i - \sum_{-r\leq\ell\leq 0} X_\ell a_{\ell i}\right) P_j + \left(a_j - \sum_{-r\leq\ell\leq 0} X_\ell a_{\ell j}\right) P_i - a_{ij} - a_{ji}$$

$$L(P_i\psi^j) = P_iL\psi^j + a_i\psi^j - \sum_{-r\leq\ell\leq 0} \left[X_\ell(a_{\ell i}\psi^j) + a_{i\ell}X_\ell\psi^j\right]$$

we have

$$L\left[(P_i+\psi^i)(P_j+\psi^j)\right] = -a_{ij}-a_{ji}+a_i\psi^j+a_j\psi^i+L(\psi^i\psi^j)$$
$$-\sum_{-r\leq \ell\leq 0}\left[X_\ell(a_{\ell i}\psi^j)+a_{i\ell}X_\ell\psi^j+X_\ell(a_{\ell j}\psi^i)+a_{j\ell}X_\ell\psi^i\right].$$

Since $\langle L(\psi^i\psi^j)\rangle = 0$, we conclude that

$$\langle L(\,(P_i+\psi^i)\,(P_j+\psi^j)\,)\rangle = -2q_{ij}$$

which proves (16.6.4).

Now, if we had

$$\sum_{1\leq i,j\leq n_1} q_{ij}\xi_i\xi_j = 0$$

then by (16.6.3), we would have that $E_iu = 0$, for all $1\leq i\leq p$, i.e. the function u would be constant along the integral curves of the vector fields $E_1,...,E_p$. Since these vector fields satisfy Hörmander's condition, u would then be a constant function, and hence we would have that

$$\xi^1P_1+...+\xi^{n_1}P_{n_1} = \xi^1\psi^1+...+\xi^{n_1}\psi^{n_1}+\text{const.}$$

This is absurd because the right term is a bounded function and the left term is not bounded.

17. Proof of propositions 1.6.3 and 1.6.4 in the general case

The goal of this section is to give the proof of propositions 1.2.3 and 1.2.4 in the general case.

Given a kernel $K(x,y)$ on G we shall set

$$\|K\|_1 = \sup\{\|K(x,.)\|_1, \|K(.,y)\|_1; x,y\in G\}$$
$$\|K\|_\infty = \sup\{\|K(x,.)\|_\infty, \|K(.,y)\|_\infty; x,y\in G\}.$$

If $S(x,y)$ is another kernel then we shall denote by KS their product $KS(x,y) = \int K(x,z)S(z,y)dy$.

As explained in section 1.14.2, we shall extend the heat kernel $p_t^H(x,y)$ of the homogenised sub-Laplacian L_H and its derivatives to G.

17.1. A first lemma relating the heat kernels $p_t(x,y)$ and $p_t^H(x,y)$. The proof of propositions 1.6.3 and 1.6.4 is based on the following:

LEMMA 17.1.1. *There is a constant $c>0$ such that for all $t>0$ and $T\geq 1$*

$$\|\,(p_t-p_t^H)\,p_T^H\|_\infty \leq c\,T^{-(D+1)/2}. \tag{17.1.1}$$

PROOF. Let

$$F_t = p_tp_T^H - p_{t+T}^H - \sum_{1\leq j\leq n_2}\psi^jX_jp_{t+T}^H - \sum_{1\leq i,j\leq n_1}\psi^{ij}X_iX_jp_{t+T}^H.$$

By (12.3), there is $c>0$ such that for all $T\geq 1$ and $t\geq 0$

$$\|X_jp_{t+T}^H\|_\infty \leq c\,(t+T)^{-(D+1)/2},\quad \|X_iX_jp_{t+T}^H\|_\infty \leq c\,(t+T)^{-(D+2)/2}. \tag{17.1.2}$$

So to prove (17.1.1) it is enough to prove that there is $c > 0$ such that for all $T \geq 1$ and $t \geq 0$

$$\|F_t\|_\infty \leq c\, T^{-(D+1)/2}. \tag{17.1.3}$$

We have

$$F_t = \int_0^t p_{t-s}\left(\frac{\partial}{\partial s} + L\right) F_s ds + p_t F_0. \tag{17.1.4}$$

Since

$$F_0 = -\sum_{1\leq j\leq n_2} \psi^j X_j p_T^H - \sum_{1\leq i,j\leq n_1} \psi^{ij} X_i X_j p_T^H,$$

it follows from (17.1.2) that

$$\|F_0\|_\infty \leq c\, T^{-(D+1)/2}. \tag{17.1.5}$$

Since

$$\begin{aligned}&\left(\frac{\partial}{\partial t} + L\right) F_t \\ &\quad= \left(\frac{\partial}{\partial t} + L\right)\left(- p_{t+T}^H - \sum_{1\leq j\leq n_2} \psi^j X_j p_{t+T}^H - \sum_{1\leq i,j\leq n_1} \psi^{ij} X_i X_j p_{t+T}^H\right),\end{aligned}$$

it follows from (16.5.6) and (12.3) that

$$\| \left(\frac{\partial}{\partial t} + L\right) F_t\|_\infty \leq c(t+T)^{-(D+3)/2}. \tag{17.1.6}$$

Combining (17.1.4), (17.1.5) and (17.1.6) we have that

$$\begin{aligned}\|F_t\|_\infty \leq& \int_0^t \|p_{t-s}\|_1 \| \left(\frac{\partial}{\partial s} + L\right) F_s\|_\infty ds + \|p_t\|_1 \|F_0\|_\infty \\ \leq& c\int_0^t (s+T)^{-(D+3)/2} ds + cT^{-(D+1)/2} \\ \leq& cT^{-(D+1)/2}\end{aligned}$$

which proves (17.1.3) and the lemma follows.

17.2. Proof of proposition 1.6.3. By (2.1.1) and the local Harnack inequality (2.2.1), it is enough to prove that for all $a \geq 1$ there is $\delta > 0$ and $r_0 > 1$ such that

$$\int_{B_r(e)} p_t(x,y) dy > \delta \tag{17.2.1}$$

for all $r \geq r_0$ and $(t,x) \in (a^{-2}r^2, a^2r^2) \times B_{ar}(e)$.

We have

$$\begin{aligned}\int_{B_r(e)} p_t(x,y)dy \geq & \frac{1}{\|p_T^H\|_\infty}\int_{B_r(e)} p_t(x,y)p_T^H(y,e)dy \\ = & \frac{1}{\|p_T^H\|_\infty}\left(p_t p_T^H(x,e) - \int_{\{y\notin B_r(e)\}} p_t(x,y)p_T^H(y,e)dy\right) \\ \geq & \frac{1}{\|p_T^H\|_\infty}\left(p_{t+T}^H(x,e) - \|\left(p_t - p_t^H\right)p_T^H\|_\infty - \int_{\{y\notin B_r(e)\}} p_t(x,y)p_T^H(y,e)dy\right)\end{aligned}$$

Let us fix $a > 1$. Then, by (1.8.2), there is $c > 1$ such that for all $r, T \geq 1$

$$\inf\{p_{t+T}^H(x,e) : x \in B_{ar}(e), a^{-2}r^2 \leq t \leq a^2r^2\} \geq \frac{1}{c}(a^2r^2 + T)^{-D/2}$$

and

$$\sup\{p_T^H(x,e) : x \notin B_r(e)\} \leq cT^{-D/2}\exp\left(-\frac{r^2}{cT}\right).$$

Also, by (17.1.1) there is a constant $c' > 0$ such that for all $t, T \geq 1$

$$\|\left(p_t - p_t^H\right)p_T^H\|_\infty \leq c'\, T^{-(D+1)/2}.$$

It follows that for all $(t,x) \in (a^{-2}r^2, a^2r^2) \times B_{ar}(e)$,

$$\begin{aligned}&\int_{B_r(e)} p_t(x,y)dy \\ &\quad \geq \frac{1}{c}T^{D/2}\left(\frac{1}{c}(a^2r^2+T)^{-D/2} - c'T^{-(D+1)/2} - cT^{-D/2}\exp\left(-\frac{r^2}{cT}\right)\right)\end{aligned}$$

If $T = \varepsilon r^2$, for some $\varepsilon \in (0,1)$, then we have

$$\begin{aligned}&\int_{B_r(e)} p_t(x,y)dy \\ &\quad \geq \frac{1}{c}\varepsilon^{D/2}\left(\frac{1}{c}(a^2+\varepsilon)^{-D/2} - c\varepsilon^{-(D+1)/2}r^{-1} - c\varepsilon^{-D/2}\exp\left(-\frac{1}{c\,\varepsilon}\right)\right) \\ &\quad \geq \frac{1}{c}\left(\frac{1}{c}(a^2+1)^{-D/2} - c\varepsilon^{-1/2}r^{-1} - c\exp\left(-\frac{1}{c\,\varepsilon}\right)\right).\end{aligned}$$

The proposition follows by chosing ε small enough and r large enough.

17.3. Proof of proposition 1.6.4. It follows from the semigroup property $p_{t+s} = p_t p_s$ and the fact that for every compact neighborhood U of e

$$\int_U p_t(e,y)dy \to 1, \quad (t \to 0)$$

that (1.6.4) holds $0 < t \leq n^2$, for any fixed $n \in \mathbb{N}$. So, by (2.1.1), it is enough to prove that for all $\delta > 0$ there is $c_\delta > 1$ and $r_0 > 1$ such that

$$\int_{\{y\in G: d(e,y)\geq c_\delta r\}} p_{r^2}(e,y)dy < \delta \tag{17.3.1}$$

for all $r \geq r_0,\ x \in G$.

Let $a_1 > a_2 \geq 1$ and set

$$M = \sup\left\{\int_{B_{a_2r}(e)} p^H_{r^2}(y,z)dz : y \notin B_{a_1r}(e)\right\}.$$

We have

(17.3.2)

$$\begin{aligned}
\int_{B_{a_1r}(e)} p_{r^2}(e,y)dy &\geq \int_{B_{a_1r}(e)}\int_{B^0_{a_2r}(e)} p_{r^2}(e,y)p^H_{r^2}(y,z)dydz \\
&= \int_{B_{a_2r}(e)} p_{r^2}p^H_{r^2}(e,z)dz \\
&\quad - \int_{\{y\notin B_{a_1r}(e)\}}\int_{B_{a_2r}(e)} p_{r^2}(e,y)p^H_{r^2}(y,z)dydz \\
&\geq \int_{B_{a_2r}(e)} p_{r^2}p^H_{r^2}(e,z)dz - M \\
&= \int_{B_{a_2r}(e)} p^H_{r^2}p^H_{r^2}(e,z)dz \\
&\quad + \int_{B_{a_2r}(e)} \left(p_{r^2} - p^H_{r^2}\right)p^H_{r^2}(e,z)dz - M \\
&\geq \int_{B_{a_2r}(e)} p^H_{2r^2}(e,z)dz - \|\left(p_{r^2} - p^H_{r^2}\right)p^H_{r^2}\|_\infty |B_{a_2r}(e)| - M.
\end{aligned}$$

Now, by (17.1.1) there is $c_1 > 0$ such that for all $r \geq 1$

$$\|\left(p_{r^2} - p^H_{r^2}\right)p^H_{r^2}\|_\infty \leq c_1\, r^{-D-1}. \tag{17.3.3}$$

Also, by (1.2) and (2.1.1), there is $c_2 > 0$ such that

$$|B_r(x)| \leq cr^D, \qquad x \in G,\ r \geq 1. \tag{17.3.4}$$

Combining (17.3.2), (17.3.3) and (17.3.4) we have

$$\int_{B_{a_1r}(e)} p_{r^2}(e,y)dy \geq \int_{B_{a_2r}(e)} p^H_{2r^2}(e,z)dz - c_1c_2a_2^Dr^{-1} - M. \tag{17.3.5}$$

Now let us fix $\delta > 0$ and let us chose $a_2 \geq 1$ so that

$$\int_{B_{a_2r}(e)} p^H_{2r^2}(e,z)dz \geq 1 - \frac{\delta}{3}$$

for all $r \geq 1$. Let us chose $a_1 > a_2$ so that $M \leq \delta/3$ and $N \geq 1$ so that

$$c_1c^2a_2^DN^{-1} \leq \frac{\delta}{3}.$$

Then, it follows from (17.3.5) that

$$\int_{B_{a_1r}(e)} p_{r^2}(e,y)dy \geq 1 - \delta$$

and the proposition follows.

18. Proof of the Gaussian estimate in the general case

The goal of this section is to prove the upper Gaussian estimate (1.8.1). The proof is similar to the proof in the nilpotent case (see section 9).

The only change is that we must replace the functions ρ_k and H_k, $k \geq 1$ with the functions ρ_k^ψ and H_k^ψ, $k \geq 1$ respectively. These functions are defined below.

18.1 The functions ρ_k^ψ. Using the notation of section 1.4.7, let us consider, on the nil-shadow $(S_1)_N$ of S_1, a family functions $\rho_k, k \geq 1$ defined as in section 9.1. Since $(S_1)_N$ is identified as differentiable manifold with S_1, we can extend these functions to $S_1 \rightthreetimes M_1$ by setting $\rho_k(x,z) = \rho_k(x)$, $(x,z) \in S_1 \rightthreetimes M_1$ and then to G by setting $\rho_k(g) = \rho_k(\pi(g))$, $g \in G$, where π is the quotient map $\pi : G \to G/C \cong S_1 \rightthreetimes M_1$.

Since the correctors $\psi^j, 1 \leq j \leq n_1$ are bounded functions there is a constant $\alpha > 0$ such that

$$\| \sum_{1 \leq j \leq n_1} \psi^j X_i \rho_k \|_\infty < \alpha, \quad k \geq 1 \tag{18.1.1}$$

Let us fix such a constant $\alpha > 0$ and denote by $\rho_k^\psi, k \geq 1$ the functions defined by

$$\rho_k^\psi = \rho_k + \sum_{1 \leq j \leq n_1} \psi^j X_i \rho_k + \alpha$$

The following lemma is an immediate consequence of lemma 9.1.1 and it shows that the functions ρ_k^ψ have essentially the same behavior as the functions $\rho_k(x)$.

LEMMA 18.1.1. *For all $n \in \mathbb{N}$ and all $X \in \mathfrak{g}$ there are constants $c, \zeta \geq 1$ such that for all $k \geq 1$,*

$$\begin{aligned}
&\rho_k^\psi(x) \geq 0, \quad x \in G \\
&\rho_k^\psi(x) \leq \ c\ |x|_G \quad x \subset G \\
&\rho_k^\psi(x) = \alpha, \quad \text{for } |x|_G \leq \sqrt{k} \\
&\frac{1}{c}\ |x| \leq \rho_k^\psi(x) \leq \ c\ |x|_G \quad \text{for } |x| \geq \zeta\sqrt{k} \\
&|X\rho_k^\psi(x)| \leq c, \quad x \in G.
\end{aligned} \tag{18.1.2}$$

Also by (16.5.3), there is a constant $c > 0$ such that for all $x \in G$ and $k \geq 1$

$$|L\rho_k^\psi(x)| \leq c\left(\nabla_X^2 \rho_k(x) + \nabla_X^3 \rho_k(x)\right)$$

and hence there is a constant $c > 0$ such that for all $x \in G$ and $k \geq 1$

$$|L\rho_k^\psi(x)| \leq \frac{c}{|x|_G}. \tag{18.1.3}$$

18.2. The functions and H_k^ψ. For fixed constants $A > 0$ and $B > 0$ we consider the family of functions H_k^ψ, $k \geq 1$ defined by

$$H_k^\psi(t,x) = \exp\left(-\frac{\left(\rho_k^\psi(x) + B\sqrt{k}\right)^2}{A(k+t)}\right), \quad t \geq 0,\ x \in G.$$

LEMMA 18.2.1. *There are constants $A > 0$ and $B > 0$ such that functions H_k^ψ, $k \geq 1$ satisfy*

$$\left(\frac{\partial}{\partial t} + L\right) H_k^\psi(t,x) > 0 \tag{18.2.1}$$

for all $(t,x) \in [0,k] \times G$.

PROOF. The proof is similar to the proof of lemma 9.2.1. We have

$$\frac{\partial}{\partial t} H_k^\psi(t,x) = \frac{1}{A}\frac{1}{k+t}\frac{\left(\rho_k^\psi(x) + B\sqrt{k}\right)^2}{k+t} H_k^\psi(t,x).$$

Also, for all smooth vector fields X and Y, we have

$$\begin{aligned} YH_k^\psi(t,x) =& -\frac{1}{A}\frac{1}{k+t}2\left(\rho_k^\psi(x) + B\sqrt{k}\right) Y\rho_k^\psi(x)\ H_k(t,x) \\ XYH_k(t,x) =& -\frac{1}{A}\frac{1}{k+t}2X\rho_k^\psi(x)Y\rho_k^\psi(x)\ H_k^\psi(t,x) \\ & -\frac{1}{A}\frac{1}{k+t}2\left(\rho_k^\psi(x) + B\sqrt{k}\right) XY\rho_k^y(x)\ H_k^\psi(t,x) \\ & +\frac{1}{A^2}\frac{1}{(k+t)^2}4\left(\rho_k(x) + B\sqrt{k}\right)^2 X\rho_k^\psi(x)Y\rho_k^\psi(x)\ H_k^\psi(t,x). \end{aligned}$$

So

$$\begin{aligned} \left(\frac{\partial}{\partial t} + L\right) H_k^\psi(t,x) =& \frac{1}{A}\frac{1}{k+t}\Bigg[\frac{\left(\rho_k^\psi(x) + B\sqrt{k}\right)^2}{k+t} \\ & + 2\sum_{1\leq i\leq p}\left(E_i\rho_k^\psi(x)\right)^2 \\ & - 2\left(\rho_k^\psi(x) + B\sqrt{k}\right) L\rho_k^\psi(x) \\ & - 4\frac{1}{A}\frac{1}{k+t}\left(\rho_k^\psi(x) + B\sqrt{k}\right)^2 \sum_{1\leq i\leq p}\left(E_i\rho_k^\psi(x)\right)^2\Bigg] H_k^\psi(t,x). \end{aligned}$$

Case I: $|x|_G \leq \sqrt{k}$.
Then

$$\left(\frac{\partial}{\partial t} + L\right) H_k^\psi(t,x) = \frac{1}{A}\frac{1}{k+t}\frac{\left(\alpha + B\sqrt{k}\right)^2}{k+t} H_k^\psi(t,x) > 0.$$

Case II: $\sqrt{k} \leq |x|_G \leq \zeta\sqrt{k}$ and $0 \leq t \leq k$. Then by (18.1.2) and (18.1.3) there is a constant $c > 0$ such that

$$\begin{aligned} \left(\frac{\partial}{\partial t} + L\right) H_k^\psi(t,x) \geq& \frac{1}{A}\frac{1}{k+t}\Bigg[\frac{B^2}{2} - c - c\left(c\sqrt{k} + B\sqrt{k}\right)\frac{1}{\sqrt{k}} \\ & - c\frac{1}{A}\frac{1}{k}\left(c\sqrt{k} + B\sqrt{k}\right)^2\Bigg] H_k^\psi(t,x) \\ =& \frac{1}{A}\frac{1}{k+t}\left[\frac{B^2}{2} - c - c(c+B) - c\frac{1}{A}(c+B)^2\right] H_k(t,x). \end{aligned}$$

So, by chosing B large enough, so that

$$\frac{B^2}{4} > c + c(c+B)$$

and A large enough, so that

$$\frac{B^2}{4} > c\frac{1}{A}(c+B)^2$$

we have

$$\left(\frac{\partial}{\partial t} + L\right) H_k^{\psi}(t,x) > 0.$$

Case III: $|x|_G > \zeta\sqrt{k}$ and $0 \le t \le k$.
Then by (18.1.2) and (18.1.3) there are constants $c_1, c_2 > 0$ such that

$$\begin{aligned}
\left(\frac{\partial}{\partial t} + L\right) H_k^{\psi}(t,x) \geq & \frac{1}{A}\frac{1}{k+t}\Bigg[\frac{\left(c_1|x|_G + B\sqrt{k}\right)^2}{2k} - c_2 - c_2\left(c_2|x|_G + B\sqrt{k}\right)\frac{1}{|x|_G} \\
& - c_2\frac{1}{A}\frac{1}{k}\left(c_2|x|_G + B\sqrt{k}\right)^2\Bigg] H_k^{\psi}(t,x) \\
= & \frac{1}{A}\frac{1}{k+t}\Bigg[\frac{\left(c_1|x|_G + B\sqrt{k}\right)^2}{2k} - c_2 - c_2^2 - c_2 B\frac{\sqrt{k}}{|x|_G} \\
& - c_2\frac{1}{A}\frac{\left(c_2|x|_G + B\sqrt{k}\right)^2}{k}\Bigg] H_k^{\psi}(t,x) \\
\geq & \frac{1}{A}\frac{1}{k+t}\Bigg[\frac{\left(c_1|x|_G + B\sqrt{k}\right)^2}{2k} - c_2 - c_2^2 - c_2 B \\
& - c_2\frac{1}{A}\frac{\left(c_2|x|_G + B\sqrt{k}\right)^2}{k}\Bigg] H_k^{\psi}(t,x).
\end{aligned}$$

So, by chosing B large enough, so that

$$\frac{\left(c_1|x|_G + B\sqrt{k}\right)^2}{4k} > c_2 + c_2^2 + c_2 B$$

and A large enough, so that

$$\frac{1}{4}\left(c_1|x|_G + B\sqrt{k}\right)^2 > c_2\frac{1}{A}\left(c_2|x|_G + B\sqrt{k}\right)^2$$

we have

$$\left(\frac{\partial}{\partial t} + L\right) H_k^{\psi}(t,x) > 0$$

for all $t \in [0,k]$ and $x \in G$ and the lemma follows.

19. A Berry-Esseen estimate for the heat kernels on connected Lie groups of polynomial volume growth

Let G be a connected Lie group of polynomial volume growth and let U a compact neighborhood U of the identity element e of G. Let us also denote by ∂_z, $z \in G$ the difference operator $\partial_z f(x) = f(xz) - f(x)$ and set

$$\nabla_U f(x) = \sup_{z\in U} |\partial_z f(x)| \quad \text{and} \quad \nabla_U f = \sup_{x\in G} |\nabla_U f(x)|.$$

Let L be a centered left invariant sub-Laplacian on G. Then by (4.3) the heat kernel $p_t(x,y)$ of L satisfies

$$\|\nabla_U p_t\|_\infty \leq ct^{-(D+\gamma)/2}, \qquad t \geq 1 \tag{19.1}$$

for some $\gamma \in (0,1]$.

Let $p_t^H(x,y)$ be the heat kernel of the homogenised sub-Lplacian L_H associated to L. As we explained in section 1.14.2, we shall extend $p_t^H(x,y)$ and its derivatives to G.

In this section we shall prove the following result:

THEOREM 21.1. *Let assume that $p_t(x,y)$ satisfies (19.1) for some $\gamma \in (0,1]$. Then there is a constant $c \geq 1$ such that*

$$\|p_t - p_t^H\|_\infty \leq ct^{-(D+\gamma)/2}, \quad t \geq 1. \tag{19.2}$$

Once we have proved theorem 1.13.2, then the above inequality will hold with $\gamma = 1$.

The following two lemmas, inspired from [**Be, Saz**] (see also [**A5**]) show that the above result can be proved by induction.

LEMMA 19.2. *There are constants $a, b \geq 1$ such that for all $t, T \geq 1$*

$$\|p_t - p_t^H\|_\infty \leq a\|(p_t - p_t^H)p_T^H\|_\infty + b\sqrt{T}t^{-(D+\gamma)/2} \tag{19.3}$$

LEMMA 19.3. *There is a constant $c \geq 1$ such that if for some $n \in \mathbb{N}$*

$$\|p_t - p_t^H\|_\infty \leq At^{-(D+\gamma)/2}, \quad \textit{for all } \ t \in [1,n], \tag{19.4}$$

then

$$\|(p_t - p_t^H)p_T^H\|_\infty \leq c\left(1 + \frac{A}{\sqrt{T}}\right)t^{-(D+\gamma)/2}, \quad \textit{for all } \ t \in (n, n+1]. \tag{19.5}$$

19.1. Proof of Lemma 19.2. Let us set

$$h_t(x,y) = p_t(x,y) - p_t^H(x,y)$$

and assume that

$$-\|h_t\|_\infty = \min\{h_t(x,y), x, y \in G\}$$

(the case $\|h_t\|_\infty = \max\{h_t(x,y), x, y \in G\}$ can be treated in the same way).

Then there are $x_0, y_0 \in G$ such that

$$h_t(x_0, y_0) = -\|h_t\|_\infty$$

and hence

$$
\begin{aligned}
-\|h_t\|_\infty \leq & \int h_t(x_0,z)p_T^H(z,y_0)dz \\
& = h_t(x_0,y_0)\int_{d(y_0,z)\leq c\sqrt{T}} p_T^H(z,y_0)dz \\
& + \int_{d(y_0,z)\leq c\sqrt{T}} [h_t(x_0,z)-h_t(x_0,y_0)]p_T^H(z,y_0)dz \\
& + \int_{d(y_0,z)\geq c\sqrt{T}} h_t(x_0,z)p_T^H(z,y_0)dz \\
\leq & -\|h_t\|_\infty \int_{d(y_0,z)\leq c\sqrt{T}} p_T^H(z,y_0)dz \\
& + c\sqrt{T}\|\nabla_U h_t(x_0,.)\|_\infty \int_{d(y_0,z)\leq c\sqrt{T}} p_T^H(z,y_0)dz \\
& + \|h_t\|_\infty \int_{d(y_0,z)\geq c\sqrt{T}} p_T^H(z,y_0)dz.
\end{aligned}
$$

Hence, if

$$
\lambda = \int_{d(y_0,z)\leq c\sqrt{T}} p_T^H(z,y_0)dz
$$

then

$$
-\|(p_t-p_t^H)p_T^H\|_\infty \leq -\|p_t-p_t^H\|_\infty\lambda + c\sqrt{T}\lambda t^{-(D+\gamma)/2} + \|p_t-p_t^H\|_\infty(1-\lambda)
$$

or

$$
(2\lambda-1)\|p_t-p_t^H\|_\infty \leq \|(p_t-p_t^H)p_T^H\|_\infty + c\lambda\sqrt{T}\lambda t^{-(D+\gamma)/2}
$$

So by chosing c large enough, so that $\lambda > 1/2$, we get

$$
\|p_t-p_t^H\|_\infty \leq \frac{1}{2\lambda-1}\|(p_t-p_t^H)p_T^H\|_\infty + \frac{c\lambda}{2\lambda-1}\sqrt{T}t^{-(D+\gamma)/2}
$$

which proves the lemma.

19.2. Proof of Lemma 19.3. We shall assume that $n \geq 4$. Let us denote by ψ^{*j}, $1 \leq j \leq n_2$ and ψ^{*ij}, $1 \leq i,j \leq n_1$ the first and second order correctors associated to the adjoint L^* of L and set

$$
\begin{aligned}
W_t(x,y) &= \sum_{1\leq j\leq n_2} \psi^j(x)X_j^x p_t^H(x,y) + \sum_{1\leq i,j\leq n_1} \psi^{ij}(x)X_i^x X_j^x p_t^H(x,y) \\
W_t^*(x,y) &= \sum_{1\leq j\leq n_2} \psi^{*j}(y)X_j^y p_t^H(x,y) + \sum_{1\leq j\leq n_1} \psi^{*ij}(y)X_i^y X_j^y p_t^H(x,y)
\end{aligned}
$$

where the superscripts x and y denote differentiation with respect to the x and y variables respectively.

Let us fix a function $0 \leq \alpha \in C^\infty(\mathbb{R})$ satisfying $|\alpha'(t)| \leq 1$, $t \in \mathbb{R}$, $\alpha(t) = 0$, for $t \leq 0$ and $\alpha(t) = 1$, for $t \geq 2$

Let us also fix a $T \geq 1$ and set

$$
U_t^1 = p_{t+T}^H + \alpha(t)W_t p_T^H \qquad \text{and} \qquad U_t = p_t p_T^H - U_t^1
$$

To prove (19.5), it is enough to prove that there is a constant $c > 0$, such that

$$\|U_t\|_\infty \le c(1 + \frac{A}{\sqrt{T}})t^{-(D+\gamma)/2} \tag{19.2.1}$$

To this end, let

$$V_t = (\frac{\partial}{\partial t} + L)U_t.$$

Note that by (16.5.6) there is a constant $c > 0$ such that for all $t \ge 1$ and $x, y \in G$,

$$|V_t(x,y)| \le c\Big(\nabla_X \frac{\partial}{\partial t} p^H_{t+T}(x,y) + \nabla^2_X \frac{\partial}{\partial t} p^H_{t+T}(x,y) + \nabla^3_X p^H_{t+T}(x,y) + \nabla^4_X p^H_{t+T}(x,y)\Big).$$

So, by (12.3), there is a constant $c > 0$ such that for all $t \ge 2$ and $T \ge 1$

$$\|V_t\|_\infty \le c(t+T)^{-(D+3)/2}, \qquad \|V_t\|_1 \le c(t+T)^{-3/2}. \tag{19.2.2}$$

For $0 \le t \le 2$, we have

$$\|V_t\|_\infty \le c(t+T)^{-(D+1)/2}, \qquad \|V_t\|_1 \le c(t+T)^{-1/2}. \tag{19.2.3}$$

Since $U_0 = 0$, we have

$$U_t = \int_0^t p_{t-s} V_s ds.$$

By (19.2.2), we have

$$\begin{aligned} |\int_{t/2}^t p_{t-s} V_s dt| \le& \int_{t/2}^t \|p_{t-s}\|_1 \|V_s\|_\infty ds \\ \le& c \int_{t/2}^t s^{-(D+3)/2} ds \\ \le& \, ct^{-(D+1)/2} \end{aligned} \tag{19.2.4}$$

To estimate the other part of the integral let us first observe that by (19.1) and the hypothesis (19.4)

$$\begin{aligned} \|p_t - p^H_t\|_\infty \le& \|p_t - p_{n-2}\|_\infty + \|p_{n-2} - p^H_{n-2}\|_\infty + \|p^H_{n-2} - p^H_t\|_\infty \\ \le& ct^{-(D+\gamma)/2} + A(\frac{t}{n-2})^{(D+\gamma)/2} t^{-(D+\gamma)/2} + ct^{-(D+2)/2} \\ \le& cAt^{-(D+\gamma)/2} \end{aligned} \tag{19.2.5}$$

for all $t \in [n-2, n+1]$.

Next we write

$$\int_0^{t/2} p_{t-s} V_s ds = \int_0^{t/2} \left(p_{t-s} - p^H_{t-s}\right) V_s ds + \int_0^{t/2} p^H_{t-s} V_s ds. \tag{19.2.6}$$

By (19.2.5) and the hypothesis (19.4) we have

(19.2.7)

$$
\begin{aligned}
&|\int_0^{t/2} \left(p_{t-s} - p^H_{t-s}\right) V_s ds| \\
&\quad \leq \int_0^2 |p_{t-s} - p^H_{t-s}||V_s| ds + \int_2^{t/2} |p_{t-s} - p^H_{t-s}||V_s| ds \\
&\quad \leq \int_0^2 \|p_{t-s} - p^H_{t-s}\|_\infty \|V_s\|_1 ds + \int_2^{t/2} \|p_{t-s} - p^H_{t-s}\|_\infty \|V_s\|_1 ds \\
&\quad \leq cAt^{-(D+\gamma)/2} \int_0^2 (s+T)^{-1/2} ds + cA \int_2^{t/2} (t-s)^{-(D+\gamma)/2} (s+T)^{-3/2} ds \\
&\quad \leq cA \frac{1}{\sqrt{T}} t^{-(D+\gamma)/2} + cAt^{-(D+\gamma)/2} \int_2^{t/2} (s+T)^{-3/2} ds \\
&\quad \leq cA \frac{1}{\sqrt{T}} t^{-(D+\gamma)/2}.
\end{aligned}
$$

To estimate the other term in (19.2.6), we observe that

$$
\int_0^{t/2} p^H_{t-s} V_s ds = \int_0^{t/2} (p^H_{t-s} + W^*_{t-s}) V_s ds - \int_0^{t/2} W^*_{t-s} V_s ds.
$$

Then, on the one hand

(19.2.8)

$$
\begin{aligned}
|\int_0^{t/2} W^*_{t-s} V_s ds| \leq & \int_0^{t/2} \|W^*_{t-s}\|_\infty \|V_s\|_1 ds \\
\leq & c \int_0^2 (t-s)^{-(D+1)/2} (s+T)^{-1/2} ds \\
& + c \int_2^{t/2} (t-s)^{-(D+1)/2} (s+T)^{-3/2} ds \\
\leq & ct^{-(D+1)/2} \left(\int_0^2 (s+T)^{-1/2} ds + \int_2^{t/2} (s+T)^{-3/2} ds \right) \\
\leq & c \frac{1}{\sqrt{T}} t^{-(D+1)/2}
\end{aligned}
$$

and on the other hand

$$
\begin{aligned}
\int_0^{t/2} (p^H_{t-s} + W^*_{t-s}) V_s ds = & \int_0^{t/2} (p^H_{t-s} + W^*_{t-s}) \left(\frac{\partial}{\partial s} + L \right) U^1_s ds \\
= & \int_0^{t/2} \left[\left(-\frac{\partial}{\partial s} + L^* \right) (p^H_{t-s} + W^*_{t-s}) \right] U^1_s ds \\
& + (p^H_{t/2} + W^*_{t/2}) U^1_{t/2} - (p^H_t + W^*_t) U^1_0
\end{aligned}
$$

where we have set

$$
\left(-\frac{\partial}{\partial s} + L^* \right) \left(p^H_{t-s} + W^*_{t-s}\right)(x,y) = \left(-\frac{\partial}{\partial s} + L^* \right) \left(p^H_{t-s}(x,.) + W^*_{t-s}(x,.)\right)(y)
$$

Now arguing as in the proof of (19.2.2) we have

$$
\begin{aligned}
&|\int_0^{t/2}\left[\left(-\frac{\partial}{\partial s}+L^*\right)\left(p^H_{t-s}+W^*_{t-s}\right)\right]U^1_s ds| \\
&\leq \int_0^{t/2}\|\left(-\frac{\partial}{\partial s}+L^*\right)\left(p^H_{t-s}+W^*_{t-s}\right)\|_\infty\|U^1_s\|_1 ds \\
&\leq c\int_0^{t/2}(t-s)^{-(D+3)/2}ds \\
&\leq ct^{-(D+1)/2}.
\end{aligned}
\tag{19.2.9}
$$

Also

$$
\begin{aligned}
(p^H_{t/2}+W^*_{t/2})U^1_{t/2}-&(p^H_t+W^*_t)U^1_0 \\
&=(p^H_{t/2}+W^*_{t/2})(p^H_{(t/2)+T}+W_{(t/2)+T})-(p^H_t+\psi^*_t)p^H_T \\
&=p^H_{t/2}W_{(t/2)+T}+W^*_{t/2}U^1_{t/2}-W^*_t p^H_T
\end{aligned}
$$

and hence

$$
\begin{aligned}
\|(p^H_{t/2}+W^*_{t/2})U^1_{t/2}-(p^H_t+W^*_t)U^1_0\|_\infty &\leq \|p^H_{t/2}\|_1\|W_{(t/2)+T}\|_\infty \\
&+\|W^*_{t/2}\|_\infty\|U^1_{t/2}\|_1+\|W^*_t\|_\infty\|p^H_T\|_1 \\
&\leq ct^{-(D+1)/2}.
\end{aligned}
\tag{19.2.10}
$$

Summing up (19.2.4), (19.2.7), (19.2.8) (19.2.9) and (19.2.10) we get

$$
\begin{aligned}
\|U_t\|_\infty &\leq ct^{-(D+1)/2}+\frac{c}{\sqrt{T}}t^{-(D+1)/2}+cA\frac{1}{\sqrt{T}}t^{-(D+\gamma)/2} \\
&\leq c\left(1+\frac{A}{\sqrt{T}}\right)t^{-(D+\gamma)/2}
\end{aligned}
$$

which proves (19.2.1) and the lemma follows.

20. Polynomials on connected Lie groups of polynomial growth

The goal of this section is to prove the proposition 1.11.1. We shall use the notation of sections 10 and 16.

Let the monomials $P_i(t,x)$, $i=0,1,2,...$ on $(S_1)_N$ be defined as in section 10 and let us associate to those monomials and to the homogenised sub-Laplacian L_H, also in the same way as in section 10, polynomials $Q_{P_i}(t,x)$, $1=0,1,2,...$ on $(S_1)_N$ satisfying (1.9.1). As explained in section 1.11.1, we extend these polynomials and their derivatives $XY...ZQ_{P_i}$, $Z,Y,...,Z\in(\mathfrak{s}_1)_N$ to G.

Let us observe that it follows from (10.2) and (10.3) that to every polynomial $P(t,x)$ on $\mathbb{R}\times G$ we can associate another polynomial $Q(t,x)$ satisfying

$$
\begin{aligned}
\left(\frac{\partial}{\partial t}+L_H\right)Q(t,x)&=P(t,x) \\
\deg_H Q(t,x)&=\deg_H P(t,x)+2.
\end{aligned}
\tag{20.1}
$$

Let us also observe that by (16.5.2) and (16.5.6) we can take

$$
\begin{aligned}
Q^{\psi}_{P_i} &= P_i + \psi^i, \quad 1 < i \le \nu_1 \\
Q^{\psi}_{P_i} &= P_i + \sum_{1\le j\le n_2} \psi^j X_j P_i + \sum_{1\le \ell, j\le n_1} \psi^{\ell j}(z) X_\ell X_j P_i, \quad \nu_1 < i \le \nu_2.
\end{aligned} \tag{20.2}
$$

Proof of proposition 1.11.1. By (20.2) we can assume that $k \ge 3$. Then as a first approximation to $Q^{\psi}_{P_i}$ we take

$$
Q^{\psi,1}_{P_i} = Q_{P_i} + \sum_{1\le j\le n_2} \psi^j X_j Q_{P_i} + \sum_{1\le \ell,j\le n_1} \psi^{\ell j} X_\ell X_j Q_{P_i}.
$$

By (16.5.6)

$$
\left(\frac{\partial}{\partial t} + L\right) Q^{\psi,1}_{P_i} = \left(\frac{\partial}{\partial t} + L_H\right) Q_{P_i} + \sum_{0\le j\le n_{k-3}} f^{i,1}_j P_j \tag{20.3}
$$

where the $f^{i,1}_j$ are functions of type QP.

Making use of (20.1), for every $\nu_{k-4} < \mu \le \nu_{k-3}$, we consider a polynomial R_μ satisfying

$$
\begin{aligned}
\left(\frac{\partial}{\partial t} + L_H\right) R_\mu &= -P_\mu \\
\deg_H Q_\mu &= \deg_H P_\mu + 2 = k-1.
\end{aligned} \tag{20.4}
$$

We also consider functions $\phi^{i,1}_\mu$ which are of type QP and which satisfy

$$
L\phi^{i,1}_\mu = -f^{i,1}_\mu + \langle f^{i,1}_\mu \rangle, \qquad \langle \phi^{i,1}_\mu \rangle = 0. \tag{20.5}
$$

Let

$$
R^f_\mu = \langle f^{i,1}_\mu \rangle \left(R_\mu + \sum_{1\le j\le n_2} \psi^j X_j R_\mu + \sum_{1\le \ell,j\le n_1} \psi^{ij} X_\ell X_j R_\mu \right) + \phi^{i,1}_\mu P_\mu.
$$

As a second approximation to $Q^{\psi}_{P_i}$ we take the corrected polynomial

$$
Q^{\psi,2}_{P_i} = Q^{\psi,1}_{P_i} + \sum_{\nu_{k-4}<\mu\le\nu_{k-3}} R^f_\mu.
$$

This polynomial satisfies

$$
\left(\frac{\partial}{\partial t} + L\right) Q^{\psi,2}_{P_i} = \left(\frac{\partial}{\partial t} + L_H\right) Q_{P_i} + \sum_{0\le j\le n_{k-4}} f^{i,2}_j P_j,
$$

where the $f^{i,2}_j$ are again functions oftype QP.

We repeat the same procedure another $k-2$ times. The corrected polynomial $Q^{\psi,k}_{P_i}$ that we shall obtain in the end will satisfy (1.11.1) and (1.11.2).

21. A Taylor formula for the heat functions on connected Lie groups of polynomial volume growth.

The goal of this section is to prove theorem 1.11.2. One way to prove this result would be to introduce an appropriate family of dilations and proceed as in section 11 (cf. [**A2**]). But then the situation becomes much more complicated. So, we shall present a variant of that method which instead of the dilations it uses the Berry-Esseen estimate (19.2). This proof has also the advantage that it can easily be adapted to the case of convolution powers of densities on connected Lie groups and probability measures on discrete groups of polynomial growth.

21.1. A uniform approximation of an L-heat function by an L_H-heat function. In this section , we shall prove the following lemma:

LEMMA 21.1.1. *There are constants $c > 0$ and $\beta \in (0,1)$ such that for all $\Theta_1 \geq 4\Theta_2$, $\Theta_2 \geq 2$, $r \geq 1$ and all functions u satisfying*

$$\left(\frac{\partial}{\partial t} + L\right) u = 0 \quad in \quad (-\Theta_1^2 r^2, \Theta_1^2 r^2) \times B_{\Theta_1 r}(e)$$

we can associate a function u^H satisfying

$$\left(\frac{\partial}{\partial t} + L_H\right) u^H = 0 \quad in \quad (-\Theta_2^2 r^2, \Theta_2^2 r^2) \times B_{\Theta_2 r}(e)$$

as well as $\|u^H\|_\infty \leq \|u\|_\infty$ and

$$\sup\left\{|u - u^H|; (-r^2, r^2) \times B_r(e)\right\} \leq c\Theta_2^{-\beta} r^{-\beta} + ce^{-\Theta_1^2/c\Theta_2^2}. \tag{21.1.1}$$

The proof of the above result is based on the Berry-Esseen estimate (19.2) as well as the following two lemmas.

Using the notation introdused in section 3.2, let us denote by $z(t)$ the diffusion process attached to the sub-Laplacian L and by τ_r^x the stopping time

$$\tau_r^x = \inf\{t : z(t) \notin B_r(x)\}.$$

Using the Gaussian estimate (1.8.1), we can obtain the following improvement of lemma 3.2.2.

LEMMA 21.1.2. *There is a constant $c > 0$ such that for all $r, t \geq 1$*

$$P_x[\tau_r^x \leq t] \leq c \exp\left(-\frac{r^2}{ct}\right). \tag{21.1.2}$$

PROOF. The proof follows the same lines as the proof of lemma 3.2.2.

We have

$$\begin{aligned}
\int_{y \notin B_r(x)} p_t(x,y)dy =& P_x[z(t) \notin B_r(x)] \\
\geq& E^{P_x}\left[P\left(t - \tau_{2r}^x, z(\tau_{2r}^x), B_r(x)^c\right); \tau_{2r}^x \leq t\right] \\
\geq& E^{P_x}\left[P\left(t - \tau_{2r}^x, z(\tau_{2r}^x), B_r(z_{\tau_{2r}^x})\right); \tau_{2r}^x \leq t\right] \\
\geq& E^{P_x}\left[P\left(t - \tau_{2r}^x, e, B_r(e)\right); \tau_{2cr}^x \leq t\right].
\end{aligned}$$

Now, we observe that (21.1.2) is interesting only for $r^2 \geq t$ and that in that case, by (1.6.2), there is $\delta > 0$ such that

$$P\left(t - \tau_{2r}^x, e, B_r(e)\right) \geq \delta.$$

So

$$\int_{y \notin B_r(x)} p_t(x,y)dy \geq \delta P_x[\tau_{2r}^x \leq r^2].$$

Since, by (1.8.1), there is a constant $c > 0$ such that for all $r, t \geq 1$

$$\int_{y \notin B_r(x)} p_t(x,y)dy \leq c \exp\left(-\frac{r^2}{ct}\right),$$

we conclude that

$$P_x[\tau_{2r}^x \leq t] \leq \frac{1}{\delta} c \exp\left(-\frac{r^2}{ct}\right)$$

which proves the lemma .

LEMMA 21.1.3. *Let $r \geq t \geq 1$ and let u be a function satisfying*

$$\left(\frac{\partial}{\partial t} + L\right) u = 0 \quad \textit{in} \quad (-2t, 2t) \times B_{2r}(x).$$

Then

$$|u(t,x) - \int_{B_r(x)} p_t(x,y)u(0,y)dy| \leq 2\|u\|_\infty P_x[\tau_r^x \leq t]. \tag{21.1.3}$$

PROOF. We have

$$\begin{aligned} u(t,x) =& E^{P_x}[u(0,z_t); \tau_r^x > t] + E^{P_x}[u(\tau_r^x, z(\tau_r^x)); \tau_r^x \leq t] \\ =& \int_{B_r(x)} \left(p_t(x,y) - E^{P_x}\left[p_{t-\tau_r^x}(z(\tau_r^x), y); \tau_r^x \leq t\right]\right) u(0,y)dy \\ &+ E^{P_x}[u(\tau_r^x, z_{\tau_r^x}); \tau_r^x \leq t]. \end{aligned}$$

Hence

$$\begin{aligned} |u(t,x) - &\int_{B_r(x)} p_t((x,y)u(0,y)dy| \\ &\leq \|u\|_\infty \left(E^{P_x}\left[P(t - \tau_r^x, z(\tau_r^x), B_r(x)); \tau_r^x \leq t\right] + P_x[\tau_r^x \leq t]\right) \\ &\leq 2\|u\|_\infty P_x[\tau_r^x \leq t] \end{aligned}$$

and the lemma follows.

PROOF OF LEMMA 21.1.1. Let u satisfy

$$\left(\frac{\partial}{\partial t} + L\right) u = 0 \quad \text{in} \quad (-\Theta_1^2 r^2, \Theta_1^2 r^2) \times B_{\Theta_1 r}(x)$$

and let us define, for $t > -\Theta_2^2 r^2$ and $x \in G$,

$$u_1(t,x) = \int_{B_{\Theta_1 r/2}(e)} p_{t+\Theta_2^2 r^2}(x,y)u(-\Theta_2^2 r^2, y)dy$$

and

$$u^H(t,x) = \int_{B_{\Theta_1 r/2}(e)} p^H_{t+\Theta_2^2 r^2}(x,y)u(-\Theta_2^2 r^2, y)dy.$$

Now by (21.1.2) and (21.1.3)

$$\begin{aligned}\sup\{|u-u_1|; (-\Theta_2^2 r^2, \Theta_2^2 r^2) \times B_{\Theta_2 r}(e)\} &\leq 2\|u\|_\infty P_x\left[\tau^x_{\Theta_1 r/2} \leq \Theta_2^2 r^2\right] \\ &\leq c\|u\|_\infty e^{-\Theta_1^2/c\Theta_2^2}.\end{aligned} \tag{21.1.4}$$

Also by interpolating the Berry-Esseen estimate (19.2) and the Gaussian estimate (1.8.1) we have that there is $\beta \in (0,1)$ such that

$$\|p_t - p_t^H\|_1 \leq ct^{-\beta}, \qquad t \geq 1$$

It follows that

$$\sup\{|u_1 - u^H|; (-r^2, r^2) \times B_r(e)\} \leq c\|u\|_\infty \Theta_2^{-\beta} r^{-\beta} \tag{21.1.5}$$

The lemma follows by summing (21.1.4) and (21.1.5).

21.2. The iteration argument.

LEMMA 21.2.1. *For all $n \in \mathbb{N}$ and $\mu \in (0,1)$ there is $r_0 > 1$, $\Theta > 1$, and $c_n > 0$ such that for all $r \geq r_0$ and all functions u satisfying*

$$\left(\frac{\partial}{\partial t} + L\right) u = 0, \quad \textit{in} \quad (-\Theta^2 r^2, \Theta^2 r^2) \times B_{\Theta r}(e), \tag{21.2.1}$$

we have

(21.2.2)

$$\sup\left\{ |u - \sum_{0\leq i\leq \nu_n} A_i (\Theta r)^{-\deg_H P_i} Q^\psi_{P_i}| \ ; \ (r^2, r^2) \times B_r(e) \right\} < c_n \Theta^{-(n+\mu)} \|u\|_\infty,$$

where the constants A_i satisfy

$$|A_i| \leq c_n (\log \Theta)^{\deg_H P_i} \|u\|_\infty,$$

for all $0 \leq i \leq \nu_n$ and

$$\left(\frac{\partial}{\partial t} + L\right)\left(\sum_{\nu_{k-1}\leq i\leq \nu_k} A_i Q^\psi_{P_i}\right) = 0,$$

for all $1 \leq k \leq n$.

PROOF. Let us fix $n \in \mathbb{N}$, $\mu \in (0,1)$, $\Theta > 8$ and a function u satisfying (21.2.1).

Then, by lemma 21.1.1 and by taking $\Theta_1 = \Theta$ and $\Theta_2 = \Theta/\log\Theta$, there is a function u^H satisfying

$$\left(\frac{\partial}{\partial t} + L_H\right) u^H = 0, \quad \text{in} \ \left(-(\Theta/\log\Theta)^2 r^2, (\Theta/\log\Theta)^2 r^2\right) \times B_{\Theta r/\log\Theta}(e),$$

as well as

$$\|u^H\|_\infty \leq \|u\|_\infty$$

and

$$\text{(21.2.3)} \quad \sup\left\{|u-u^H| \ ; \ (-r^2,r^2)\times B_r(e)\right\} \leq c\|u\|_\infty \left(\Theta^{-\beta}(\log\Theta)^\beta r^{-\beta} + e^{-(\log\Theta)^2/c}\right).$$

Also, by theorem 11.3.2

$$\text{(21.2.4)} \quad \sup\left\{|u^H - \sum_{0\leq i\leq \nu_n} B_i\big(\frac{\Theta}{\log H}r\big)^{-\deg_H P_i} Q_{P_i}| \ ; \ (r^2,r^2)\times B_r(e)\right\} < c_n\big(\frac{\Theta}{\log H}\big)^{-(n+1)}\|u^H\|_\infty,$$

where the constants B_i satisfy

$$|B_i| \leq c_n\|u^H\|_\infty,$$

for all $0\leq i\leq \nu_n$ and

$$\left(\frac{\partial}{\partial t}+L_H\right)\left(\sum_{\nu_{k-1}\leq i\leq \nu_k} B_i Q_{P_i}(t,x)\right) = 0,$$

for all $1\leq k\leq n$.

Now let us observe that there is a constant $c>0$ such that

$$\sup\left\{|Q_{P_i} - Q^\psi_{P_i}| \ ; \ (-r^2,r^2)\times B_r(e)\right\} \leq c r^{\deg_H P_i - 1}$$

and hence

$$\text{(21.2.5)} \quad \left(\frac{\Theta}{\log\Theta}r\right)^{-\deg_H P_i} \sup\left\{|Q_{P_i} - Q^\psi_{P_i}| \ ; \ (-r^2,r^2)\times B_r(e)\right\} \leq c\left(\frac{\Theta}{\log\Theta}\right)^{-\deg_H P_i} r^{-1},$$

for all $r\geq 1$ and $i=1,2,\dots.\nu_n$.

Let us take

$$A_i = B_i(\log\Theta)^{\deg_H P_i}, \qquad 0\leq i\leq \nu_n.$$

Then, combining (21.2.3), (21.2.4) and (21.2.5), we have that

$$\begin{aligned}
&\sup\Big\{|u - \sum_{0\le i\le \nu_n} A_i\,(\Theta r)^{-\deg_H P_i} Q^{\psi}_{P_i}| \;;\; (r^2, r^2)\times B_r(e)\Big\}\\
&\le \sup\Big\{|u^H - \sum_{0\le i\le \nu_n} B_i(\log\Theta)^{\deg_H P_i}(\Theta r)^{-\deg_H P_i} Q_{P_i}| \;;\; (r^2, r^2)\times B_r(e)\Big\}\\
&\quad + \sup\big\{|u - u^H| \;;\; (-r^2, r^2)\times B_r(e)\big\}\\
&\quad + \sum_{0\le i\le \nu_n} |B_i|(\log\Theta)^{\deg_H P_i}(\Theta r)^{-\deg_H P_i} \sup\Big\{|Q_{P_i} - Q^{\psi}_{P_i}| \;;\; (r^2, r^2)\times B_r(e)\Big\}\\
&\le c_n\Theta^{-(n+1)}(\log\Theta)^{n+1}\|u^H\|_\infty + c\|u\|_\infty\Big((\log\Theta)^{\beta}\Theta^{-\beta}r^{-\beta} + e^{-(\log\Theta)^2/c}\Big)\\
&\quad + c(\log\Theta)^n\Theta^{-1}r^{-1}\sum_{0\le i\le \nu_n}|B_i|\\
&\le c_n\Theta^{-(n+1)}(\log\Theta)^{n+1}\|u\|_\infty + c\|u\|_\infty\Big((\log\Theta)^{\beta}\Theta^{-\beta}r^{-\beta} + e^{-(\log\Theta)^2/c}\Big)\\
&\quad + c(\log\Theta)^n\Theta^{-1}r^{-1}c_n\nu_n\|u\|_\infty.
\end{aligned}$$

The lemma follows by taking Θ and r_0 large enough.

LEMMA 21.2.2. *Let μ, Θ and r_0 be as in the previous lemma. Then there is a constant $c_n > 0$ such that for all $m \in \mathbb{N}$, all $r \ge \Theta^{m-1}r_0$ and all functions u satisfying*

$$\left(\frac{\partial}{\partial t} + L\right) u = 0, \quad in \quad (-\Theta^{2m}r^2, \Theta^{2m}r^2)\times B_{\Theta^m r}, \tag{21.2.6}$$

we have

$$\sup\left\{|u - \sum_{0\le i\le \nu_n} A_i^m\,(\Theta^m r)^{-\deg_H P_i} Q^{\psi}_{P_i}| \;;\; (r^2, r^2)\times B_r(e)\right\} < \Theta^{-m(n+\mu)}\|u\|_\infty, \tag{21.2.7}$$

where the constants A_i^m satisfy

$$|A_i^m| \le c_n(\log\Theta)^{-\deg_H P_i}\|u\|_\infty,$$

for all $0 \le i \le \nu_n$ and

$$\left(\frac{\partial}{\partial t} + L\right)\left(\sum_{\nu_{k-1}< i\le \nu_n} A_i^m Q^{\psi}_{P_i}\right) = 0,$$

for all $1 \le k \le n$.

PROOF. We shall prove the lemma by induction on m. For $m = 1$ we are in the case of the previous lemma. So let us assume that (21.2.7) is true for some

$m \in \mathbb{N}$. To prove that it is also true for $m+1$, let us assume, for simplicity, that $\|u\|_\infty \leq 1$. By the induction hypothesis

$$\sup\left\{ |u - \sum_{0\leq i\leq \nu_n} A_i^m \left(\Theta^{m+1} r\right)^{-\deg_H P_i} Q_{P_i}^\psi| \ ; \ (-\Theta^2 r^2, \Theta^2 r^2) \times B_{\Theta r}(e) \right\} \leq \Theta^{-m(n+\mu)}.$$

We consider the function

$$w = \Theta^{m(n+\mu)} \left(u - \sum_{0\leq i\leq \nu_n} A_i^m \left(\Theta^{m+1} r\right)^{-\deg_H P_i} Q_{P_i}^\psi \right).$$

Then

$$\left(\frac{\partial}{\partial t} + L\right) w = 0, \quad \text{in} \quad (-\Theta^2 r^2, \Theta^2 r^2) \times B_{\Theta r}(e)$$

and $\sup\left\{|w| \ ; \ (-\Theta^2 r^2, \Theta^2 r^2) \times B_{\Theta r}(e)\right\} \leq 1$. So, by lemma 21.2.1, we have that

$$\sup\left\{ |w - \sum_{0\leq i\leq \nu_n} B_i (\Theta r)^{-\deg_H P_i} Q_{P_i}^\psi| \ ; \ (r^2, r^2) \times B_r(e) \right\} < \Theta^{-(n+\mu)},$$

where the constants B_i satisfy

$$|B_i| \leq c_n (\log \Theta)^{\deg_H P_i},$$

for all $0 \leq i \leq \nu_n$ and

$$\left(\frac{\partial}{\partial t} + L\right) \left(\sum_{\nu_{k-1} < i \leq \nu_k} B_i Q_{P_i}^\psi \right) = 0,$$

for all $1 \leq k \leq n$.

So, if we set

$$A_i^{m+1} = A_i^m + \Theta^{-m(n+\mu-\deg_H P_i)} B_i,$$

then we have

$$\sup\left\{ |u - \sum_{0\leq i\leq \nu_n} A_i^{m+1} \left(\Theta^{m+1} r\right)^{-\deg_H P_i} Q_{P_i}^\psi| \ ; \ (r^2, r^2) \times B_r(e) \right\} < \Theta^{-(m+1)(n+\mu)},$$

which proves the inductive step and the lemma follows.

21.3. A Taylor formula for the heat functions. Let us observe that if $R \geq r \geq r_0$ then $\Theta^{m-1} r \leq R < \Theta^m r$ for some $m \in \mathbb{N}$. So, as a consequence of lemma 21.2.2 we have the following result (which contains theorem 1.11.2):

THEOREM 21.3.1. *For all $n \in \mathbb{N}$ there is there is a constant $c_n > 0$ such that for all $R \geq r \geq 1$ and all functions u satisfying*

$$\left(\frac{\partial}{\partial t} + L\right) u = 0, \quad in \ (-R^2, R^2) \times B_R,$$

we have

$$\sup\left\{|u - \sum_{0 \leq i \leq \nu_n} A_i R^{-\deg_H P_i} Q^{\psi}_{P_i}| \ ; \ (r^2, r^2) \times B_r(e)\right\} < c_n \left(\frac{R}{r}\right)^{-(n+1)} \|u\|_\infty, \tag{21.3.1}$$

where the constants A_i satisfy

$$|A_i| \leq c_n \|u\|_\infty,$$

for all $0 \leq i \leq \nu_n$ and

$$\left(\frac{\partial}{\partial t} + L\right)\left(\sum_{\nu_{k-1} < i \leq \nu_k} A_i Q^{\psi}_{P_i}\right) = 0,$$

for all $1 \leq k \leq n$.

22. Harnack inequalities for the derivatives of the heat functions

As an immediate consequence of theorem 21.3.1 and the local Harnack inequality (2.2.1) we have that for all $k \in \mathbb{N}$ and every $X \in \mathfrak{g}$ there is a constant $c_k > 0$ such that for all $r \geq 1$ and every function u satisfying $\left(\frac{\partial}{\partial t} + L\right) u = 0$ in $\left(-r^2, r^2\right) \times B_r(e)$ we have

$$|\frac{\partial^k}{\partial t^k} X u(0, e)| \leq c_k r^{-2k-1} \|u\|_\infty. \tag{22.1}$$

The above inequality, combined with (3.1), implies the following result (which contains theorem 1.13.2):

THEOREM 22.1. *For all $a, b > 0$ and all $k, n \in \mathbb{N}$ and every $X \in \mathfrak{g}$ there is $\alpha, \beta > 1, \alpha < \beta$, $c > 0$ and $c_k > 0$ such that for all $r \geq 1$ and all $u \geq 0$ satisfying*

$$\left(\frac{\partial}{\partial t} + L\right) u = 0 \quad in \quad \left(0, (\beta + b^2) r^2\right) \times B_{cr}(e)$$

we have

$$\sup\left\{|\frac{\partial^k}{\partial t^k} X u| \ ; \ \left(\alpha r^2, (\alpha + a^2) r^1\right) \times B_{ar}(e)\right\} \leq c_k r^{-2k-1} \inf\left\{u; \left(\beta r^2, (\beta + b^2) r^2\right) \times B_{br}(e)\right\}. \tag{22.1}$$

Applying the above result to the heat kernel $p_t(x, y)$ of L we have the following:

COROLLARY 22.2. *For all $k \in \mathbb{N}$ and every $X \in \mathfrak{g}$ there is $c > 0$ such that*

$$|\frac{\partial^k}{\partial t^k} X p_t(x, y)| \leq c t^{-(D/2)-k-(1/2)} \exp\left(-\frac{|x^{-1} y|_G}{ct}\right) \tag{22.2}$$

for all $t \geq 1$ and $x, y \in G$.

23. Harmonic functions of polynomial growth

In this section we shall give the proof of theorem 1.12.1. The approach is the same as in section 13.

PROOR OF THEOREM 1.12.1. Let u be an L-harmonic function on G which grows polynomially. Then, by (2.1.2), there are constants $c > 0$ and $n \in \mathbb{N}$ such that

$$\sup \|u\| \; ; \; B_r(e)\} \leq cr^n, \quad r \geq 1. \tag{23.1}$$

By theorem 21.3.1, there is a constant $c > 0$ such that for all $r \geq 1$

$$\sup\left\{ |u - \sum_{0\leq i\leq \nu_n} A_i^r r^{-\deg_H P_i} Q^{\psi}_{P_i}|; B_1(e) \right\} < c_n r^{-(n+1)} \|u\|_{L^\infty(B_r(e))}, \tag{23.2}$$

where we have set $Q^{\psi}_{P_i}(t,x) = Q^{\psi}_{P_i}(0,x)$ and where the constants A_i^r are such that

$$L\left(\sum_{0\leq i\leq \nu_n} A_i^r r^{-\deg_H P_i} Q^{\psi}_{P_i} \right) = 0.$$

We set $C_{r,i} = 0$, for $Q^{\psi}_{P_i} = 0$ and

$$C_{r,i} = A_i^r r^{-\deg_H P_i}$$

otherwise. Then (23.1) and (23.2) imply that there is $c > 0$ such that

$$\sup\left\{ |u - \sum_{0\leq i\leq \nu_n} C_{r,i} Q^{\psi}_{P_i}|; B_1(e) \right\} < c_n r^{-1}, \tag{25.3}$$

with

$$L\left(\sum_{0\leq i\leq \nu_n} C_{r,i} Q^{\psi}_{P_i} \right) = 0. \tag{23.4}$$

Now, there is a subsequence r_j, $r_j \to \infty$ $(j \to \infty)$, and constants C_i such that

$$C_{r_j,i} \to C_i \quad \text{as } j \to \infty \tag{23.5}$$

for all $0 \leq i \leq \nu_n$.

To see this, let us observe that if this was not the case then we would have that

$$M_r = \max\{|C_{r,i}|, \; 0 \leq i \leq \nu_n\} \to \infty \quad \text{as } r \to \infty.$$

Since $|C^{r,i}|/M_r \leq 1$, there is a subsequence r_k, $r_k \to \infty$ $(k \to \infty)$ and constants B_i such that

$$\frac{C^{r_k,i}}{M_{r_k}} \to B_i \quad \text{as } k \to \infty$$

for all $0 \leq i \leq \nu_n$. Note that the subsequence r_k can be chosen is such a way that some of the constants B_i are equal to 1 (and hence not all of them vanish). Let

$$R(x) = \sum_{0\leq i\leq \nu_n} B_i Q^{\psi}_{P_i}(x).$$

By (23.3)

$$\frac{1}{M_{r_k}}\left(u(x) - \sum_{0\le i\le \nu_n} C_{r_k,i} Q^{\psi}_{P_i}(x)\right) \to 0 \quad \text{as } k \to \infty$$

for all $x \in B_1(e)$. So, $R(x) = 0$ for all $x \in B_1(e)$. Since $LR(x) = 0$, we also have that $R(x) = 0$ for all $x \in G$.

Since, by (1.11.1)

$$\sup\{|R(x)|; B_r(e)\} \sim \sup\{|\sum_{\nu_{n-1}<i\le\nu_n} B_i P_i(x)|; B_r(e)\} \qquad \text{as } r \to \infty,$$

we have that

$$\sum_{\nu_{n-1}<i\le\nu_n} B_i P_i(x) = 0, \qquad x \in G$$

and hence $B_i = 0$, $\nu_{n-1} < i \le \nu_n$.

Arguing in the same way, we can prove successively that for all $k = n-1, ..., 1$, $B_i = 0, \nu_{k-1} < i \le \nu_k$ and that $B_0 = 0$. This is absurd because, by construction, not all of the coefficients B_i vanish.

We conclude therefore that (23.5) holds. By letting $j \to \infty$, it follows from (23.3) and (23.4) that

$$u(x) = \sum_{0\le i\le\nu_n} C_i Q_{P_i}(x)$$

for all $x \in B_1(x)$ and hence for all $x \in G$ (cf. [**Bo,** corollary 4.1]).

24. Berry-Esseen type of estimates for the derivatives of the heat kernel

The goal of this section is to prove the Berry-Esseen estimates (1.14.8) and (1.14.9).

We shall use the notation of section 19.2. We set

$$U_t(x,y) = p_t(x,y) - p_t^H(x,y) - \sum_{1\le j\le n_2} \psi^j(x)X_j^x p_t^H(x,y) - \sum_{1\le i,j\le n_1} \psi^{ij}(x)X_i^x X_j^x p_t^H(x,y).$$

Then by (12.3) and (1.14.5) there is $c>0$ such that

$$\|U_t\|_\infty \le ct^{-(D+1)/2}, \quad t\ge 1. \tag{26.1}$$

24.1. Proof of theorem 1.14.9. It is enough to prove that there is a constant $c>0$ such that

$$\|YU_t\|_\infty \le ct^{-(D+2)/2}, \quad t\ge 1. \tag{24.1.1}$$

To this end, let

$$V_t = (\frac{\partial}{\partial t} + L)U_t.$$

Then

$$U_t = \int_{t/2}^t p_{t-s}V_s ds + p_{t/2}U_{t/2}$$

and hence

$$YU_t = \int_{t/2}^t (Yp_{t-s})V_s ds + \int (Yp_{t/2})U_{t/2}.$$

Now, by (26.1)

$$\|U_{t/2}\|_\infty \le ct^{-(D+1)/2}.$$

Also, by (12.3) and (16.5.6)

$$\|V_t\|_\infty \le ct^{-(D+3)/2}, \quad t\ge 1.$$

So,

$$\begin{aligned}\|YU_t\|_\infty \le & \int_{t/2}^t \|Yp_{t-s}\|_1\|V_s\|_\infty ds + \|Yp_{t/2}\|_1\|U_{t/2}\|_\infty \\ \le & c\int_{t/2}^t (t-s)^{-1/2}s^{-(D+3)/2}ds + c(\frac{t}{2})^{-1/2}(\frac{t}{2})^{-(D+1)/2} \\ \le & ct^{\frac12}t^{-(D+3)/2} + ct^{-1/2}t^{-(D+1)/2} \\ \le & ct^{-(D+2)/2}\end{aligned}$$

which proves (24.1.1) and the theorem follows.

24.2. Proof of theorem 1.14.8. It is enough to prove that there is a constant $c>0$ such that

$$\|\frac{\partial}{\partial t}U_t\|_\infty \le ct^{-(D+3)/2}, \quad t\ge 1 \tag{24.2.1}$$

As in the proof of theorem 1.14.9, we set

$$V_t = (\frac{\partial}{\partial t} + L)\frac{\partial}{\partial t}U_t.$$

Then

$$\frac{\partial}{\partial t}U_t = \int_{t/2}^{t} p_{t-s}V_s ds + p_{t/2}\frac{\partial}{\partial t}U_{t/2} \tag{24.2.2}$$

Also, by (12.3) and (16.5.6)

$$\|V_t\|_\infty \le ct^{-(D+5)/2}, \quad t \ge 1.$$

We have

$$\begin{aligned} |\int_{t/2}^{t} p_{t-s}V_s ds| \le & \int_{t/2}^{t} \|p_{t-s}\|_1 \|V_s\|_\infty ds \\ \le & c\int_{t/2}^{t} s^{-(D+5)/2} ds \\ \le & ct^{-(D+3)/2}. \end{aligned} \tag{24.2.3}$$

Also

(24.2.4)

$$\begin{aligned} |p_{t/2}\frac{\partial}{\partial t}U_{t/2}(x,y)| =& |p_{t/2}\left(\frac{\partial}{\partial t}+L\right)U_{t/2}(x,y) - p_{t/2}LU_{t/2}(x,y)| \\ \le & \|p_{t/2}\|_1 \|\left(\frac{\partial}{\partial t}+L\right)U_{t/2}\|_\infty + |\left\langle L^* p_{t/2}(x,.), U_{t/2}(.,y)\right\rangle| \\ \le & ct^{-(D+3)/2} + \left\langle \frac{\partial}{\partial t}p_{t/2}(x,.), U_{t/2}(.,y)\right\rangle \\ \le & ct^{-(D+3)/2} + \|\frac{\partial}{\partial t}p_{t/2}\|_1 \|U_{t/2}\|_\infty \\ \le & ct^{-(D+3)/2} + ct^{-1}t^{-(D+1)/2} \\ \le & ct^{-(D+3)/2}. \end{aligned}$$

Combining (24.2.2), (24.2.3) and (24.2.4) we have (24.2.1) and the theorem follows.

25. Riesz transforms

25.1. Riesz transforms at infinity on nilpotent Lie groups. In this section we shall give the proof of theorem 1.15.4. We start with the observation that the kernel $K_{n,\infty}$ of the operator $R_{n,\infty}$ is given by

$$K_{-n/2,\infty}(x,y) = \frac{1}{\Gamma(n/2)}\int_1^\infty t^{(n/2)-1}Y_1...Y_n p_t(x,y)dt.$$

By the Gaussian estimate (12.3) we have that $K_{-n/2,\infty}$ satisfies the standard estimates

$$\begin{aligned} |K_{-n/2,\infty}(x,y)| \le & \frac{c}{|y^{-1}x|_G^D} \\ |\nabla_X^x K_{-n/2,\infty}(x,y)| + |\nabla_X^y K_{-n/2,\infty}(x,y)| \le & \frac{c}{|y^{-1}x|_G^{D+1}} \end{aligned} \tag{25.1.1}$$

for all $x, y \in G$ and where the superindices x and y denote differentiation with respect to the variables x and y respectively.

So by the Calderon-Zygmund theory (cf. [**CW, S1**]), to prove theorem 1.15.4, it is enough to prove that the operator $R_{n,\infty}$ is bounded on L^2. This can be done by an almost orthogonality argument (cf. [**S3** ch. VII]).

Let us denote by $T_j,\ j \in \mathbb{N}$ the operators with kernel K_j given by

$$K_j(x,y) = \frac{1}{\Gamma(n/2)} \int_{2^{j-1}}^{2^j} t^{(n/2)-1} Y_1...Y_n p_t(x,y)dt.$$

Then $R_{n,\infty} = \sum_{j\geq 1} T_j$. Also

$$\int K_j(x,y)dy = \int K_j(x,y)dx = 0. \tag{25.1.2}$$

By (1.13.1), there is a constant $c > 1$ such that

$$|Y_1...Y_n p_t(x,y)| \leq ct^{-n/2} p_{ct}(x,y), \quad x,y \in G, t \geq 1.$$

Hence,

$$\sup_{j\in\mathbb{N}} \|T_j\|_{L^2\to L^2} < \infty. \tag{25.1.3}$$

Finally, it follows from (12.3) by a straightforward calculation that there is a constant $c > 0$ such that for all $j \in \mathbb{N}$ and $x \in G$

$$\begin{gathered}\int |x|_G |K_j(x,y)|dx \leq c2^{j/2}, \qquad \int |y|_G |K_j(x,y)|dy \leq c2^{j/2}, \\ \int |K_j(x,y) - K_j(e,y)|dy \leq c2^{-j/2}|x|_G \\ \int |K_j(x,e) - K_j(x,y)|dx \leq c2^{-j/2}|y|_G\end{gathered} \tag{25.1.4}$$

It follows from (25.1.2), (25.1.3) and (25.1.4) that there is a constant $c > 0$ such that

$$\|T_i T_j^*\|_{L^2\to L^2} + \|T_i^* T_j\|_{L^2\to L^2} \leq c2^{-|i-j|/2}$$

and from this we conclude that $R_{n,\infty}$ is bounded on L^2 (for details we refer the reader to [**S3** pp. 623-625]).

The same arguments also apply for the operator $R_{n,\infty}^*$.

25.2. Riesz transforms at infinity on Lie groups of polynomial growth. If G is a general connected Lie group of polynomial growth, then the kernel K_∞ of the Riesz trasform R_∞ does not satisfy the standard estimates (25.1.1). So to prove theorem 1.15.3 we shall use (1.14.9) and theorem 1.15.4.

More presicely, using the notation of section 1.14, let us consider the kernels

$$K_\infty^H(x,y) = \frac{1}{\Gamma(1/2)} \int_1^\infty t^{-1/2} Y p_t^H(x,y)dt$$

$$K_{j,\infty}^H(x,y) = \frac{1}{\Gamma(1/2)} \int_1^\infty t^{-1/2} X_j p_t^H(x,y)dt, \quad 1 \leq j \leq n_1$$

and denote by R_∞^H and $R_{j,\infty}^H$, $1 \leq j \leq n_1$ respectively the associated operators.

Let us also cosider the kernel

$$S_i(x,y) = K_\infty(x,y) - K_\infty^H(x,y) - \sum_{1\leq j\leq n_1} (Y\psi^j(x)) K_{j,\infty}^H(x,y), \quad x,y \in G.$$

Then it follows from (1.14.9) that for all $\varepsilon \in (0,1)$ there is a constant $c > 0$ such that

$$|S(x,y)| \leq \frac{c}{|y^{-1}x|_G^{D+\varepsilon}}, \quad x, y \in G$$

i.e. the kernel $S_i(x,y)$ is integrable and hence the operator

$$S = R_\infty - R_\infty^H - \sum_{1 \leq j \leq n_1} (Y\psi^j)\, R_{j,\infty}^H$$

is bounded on L^p, $1 \leq p \leq \infty$.

Since, by theorem 1.15.4 the Riesz trasforms R_∞^H and $R_{j,\infty}^H$, $1 \leq j \leq n_1$ are bounded on L^p, $1 < r < \infty$ and from L^1 to weak-L^1, the same will also be true for the operators R_∞.

The same arguments also apply to the operator R_∞^*.

25.3 Local Riesz transforms. The theorem 1.15.2 can be proved by the same method as theorem 1.15.4. The estimates we need can be proved by using the local Harnack inequality 2.2.2. We omit the details.

Bibliography

[A1] G.K.Alexopoulos, *An application of Homogenization theory to Harmonic analysis on solvable Lie groups of polynomial growth*, Pacific J. Math. **159** (1993), 19-45.

[A2] G.K.Alexopoulos, *An application of homogenisation theory to harmonic analysis: Harnack inequalities and Riesz tranforms on Lie groups of polynomial growth*, Canadian J. of Math. **44** (1992), 1-37.

[A3] G.K.Alexopoulos, *Puissances de convolution sur les groupes à croissance polynomiale du volume*, C. R. Acad. Sci. Paris Serie I **324** (1997), 771-776.

[A4] G.K.Alexopoulos, *Sous-laplaciens et densités centrées sur les groupes de Lie à croicansse polynomiale du volume*, C. R. Acad. Sci. Paris Serie I **326** (1998), 539-542.

[A5] G.K.Alexopoulos, *On the large time behavior of the heat kernels of quasiperiodic differential operators*, J. Geom. Anal. (to appear).

[ALo] G.K.Alexopoulos and N.Lohoué, *Sobolev inequalities and harmonic functions of polynomial growth*, J. London Math. Soc. **48** (1993), 452-464.

[Ar] D.C.Aronson, *Bounds for the fundamental solution of a parabolic equation*, Bull. Amer. Math. Soc. **73** (1967), 890-896.

[AL1] M.Avellaneda and F.H.Lin, *Compactness methods in the theory of Homogenisation*, Comm. Pure Appl. Math. **40** (1987), 803-847.

[AL2] M.Avellaneda and F.H.Lin, *Un théorème de Liouville pour des equations elliptiques avec coefficients périodiques*, C. R. Acad. Sci. Paris Serie I **609** (1989), 245-250.

[Au1] L.Auslander, *An exposition of the structure of solvmanifolds. Part I: Algebraic theory*, Bull. Amer. Math. Soc. **79** (1973), 227-261.

[Au2] L.Auslander, *An exposition of the structure of solvmanifolds. Part II: G-induced flows*, Bull. Amer. Math. Soc. **79** (1973), 262-285.

[AG] L.Auslander and L.W.Green, *G-induced flows*, Amer. J. Math. **88** (1966), 43-60.

[BL1] G.Ben Arous et R.Léandre, *Décroissance exponentielle du noyau de la chaleur sur la diagonale (I)*, Probab. Th. Rel. Fields **90** (1991), 175-202.

[BL2] G.Ben Arous et R.Léandre, *Décroissance exponentielle du noyau de la chaleur sur la diagonale (II)*, Probab. Th. Rel. Fields **90** (1991), 377-402.

[BLP] A.Bensoussan, J.L.Lions and G.Papanicolaou, *Asymptotic Analysis of Periodic Structures*, North Holland Publ., 1978.

[Be] H.Bergström, *On the Central Limit Theorem in* $\mathbb{R}^k$, Z. Wahr. verw. Geb. **14** (1969), 113-126.

[Bo] J.M.Bony, *Principe du maximum, inégalité de Harnack et unicité du problème de Cauchy pour les opéreteurs elliptiques dégénérés*, Ann. Inst. Fourier Grenoble **19** (1969), 277-304.

[Bou1] N.Bourbaki, *Eléments de mathématiques, Livre VI, Intégration*, Hermann, Paris, 1963.

[Bou2] N.Bourbaki, *Groupes et algèbres de Lie, Chapitres 7 et 8*, Eléments de mathématique, Hermann, Paris, 1975.

[CG] M.Christ and D.Geller, *Singular integral characterizations of Hardy spaces on homogeneous groups*, Duke Math. J. **51** (1984), 547-598.

[CW] R.Coifman and G.Weiss, *Analyse non-commutetive sur certains espaces homogènes*, Lecture Notes in Mathematices, vol. 242.

[CM] T.H.Colding and W.P.Minicozzi II, *Harmonic functions with polynomial growth*, J. Diff. Geom. **46** (1997), 1-77.

[Fe] W.Feller, *An introduction to the Theory of Probability and its Applications*, vol. II, Wiley, 1971.

[Fo] G.B.Folland, *Subelliptic estimates and function spaces on nilpotent Lie groups*, Ark. f. Mat. **13** (1975), 161-207.

[FS] G.B.Folland and E.Stein, *Hardy spaces on Homogeneous groups*, Princeton University Press, 1982.

[Go] V.V.Gorbatsevich, *Splitting theorems of Lie groups and their application to the study of homogeneous spaces*, Math. USSR Izvestija **15** (1980), 441-467.

[GOV] V.V.Gorbatsevich, A.L.Onishchik and E.B.Vinberg, *Structure of Lie Groups and Lie Algebras*, Lie Groups and Lie Algebras III (A.L.Onishchik and E.B.Vinberg, eds.), Encyclopaedia of Mathematical Sciences, Springer-Verlag, 1994.

[Gui] Y.Guivarc'h, *Croissance polynômiale et périodes de fonctions harmoniques*, Bull. Sc. Math. France **101** (1973), 149-152.

[He] W.Hebish, *Estimates on the semigroups generated by left invariant operators on Lie groups*, J. reine angew. Math. **423** (1992), 1-45.

[HS] W.Hebish and L.Saloff-Coste, *Gaussian estimates for Markov chains and random walks on groups*, Ann. of Proba. **21** (1993), 673-709.

[Ho] G.Hochschild, *The Structure of Lie Groups*, Holden-Day, Inc., 1965.

[HM] G.Hochschild and G.D.Mostow, *Extensions and representations of Lie groups and Lie algebras I*, Amer. J. Math. **79** (1957), 924-942.

[Hö] L.Hörmander, *Hypoelliptic second order differential operators*, Acta. Math. **16** (1967), 147-171.

[Hu] G.A.Hunt, *Semi-groups of measures on Lie groups*, Trans. Amer. Math. Soc. **81** (1956), 264-293.

[J] N.Jacobson, *Lie Algebras*, Wiley, 1962.

[JKO] V.V.Jikov, S.M.Kozlov and O.A.Oleinik, *Homogeneization of Differential Operators and Integral Functionals*, Springer-Verlag, 1994.

[Ke] C.E.Kenig, *Harmonic Analysis Techniques for Second Order Elliptic Boundary Value Problems*, CBMS Regional Conference Series in Mathematics, vol. 83.

[Ko1] S.M.Kozlov, *Averaging differential equations with almost periodic rapidly oscillating coefficients*, Matem. Sbornik **107** (1978), no. 2, 199-217; English trasl., Math. USSR, Sbornik **35** (1979), no. 4, 481-498.

[Ko2] S.M.Kozlov, *Asymptotics of fundamental solutions for second order differential equations*, Matem. Sbornik **113:2 (155)** (1980), 302-323; English trasl., Math. USSR, Sbornik **41** (1982), no. 2, 249-267.

[KS] N.V.Krylov and M.V.Safonov, *A certain property of solutions of parabolic equations with measurable coefficients*, Math. USSR-Izs. **16** (1981), 151-164.

[Law] G.F.Lawler, *Estimates for differences and Harnack inequality for difference operators coming from random walks with symmetric, spacially inhomogeneous, increments*, Proc. London Math. Soc. **63** (1991), no. 3, 552-568.

[Lo] N.Lohoué, *Comparaison des champs de vecteurs et des puissances du Laplacien sur une variété Riemmannienne à courbure non positive*, J. Funct. Analysis **61** (1983), 164-201.

[LV] N.Lohoué and N.Th.Varopoulos, *Remarques sur les transformés de Riesz sur les groupes de Lie nilpotents*, C. R.A cad. Paris (I) 11 **301** (1985).

[M1] J.Moser, *On Harnack theorem for elliptic differential equations*, Comm. Pure Appl. Math. **14** (1961), 47-79.

[M2] J.Moser, *On Harnack theorem for elliptic differential equations*, Comm. Pure Appl. Math. **17** (1964), 101-134; **20** (1967), 232-236.

[MV] S.Moustapha and N.Th.Varopoulos, *Book to appear*, Cambridge University Press.

[NRS] A.Nagel, F.Ricci and E.Stein, *Harmonic Analysis and fundamental solutions on nilpotent Lie groups*, Analysis and Partial Differential Equations: A collection of papers dedicated to M.Cotlar (C.Sadosky, ed.), Lecture Notes in Pure and Applied Mathematics, vol. 122, Marcel Dekker, Inc., 1990, pp. 249-275.

[NSW] A.Nagel, E.Stein and M.Wainger, *Balls and metrics defined by vector fields*, Acta Math. **155** (1985), 103-147.

[P] N.V.Pedersen, *On the basic algebraic operations in solvable Lie groups*, Aequationes math. **48** (1994), 228-253.

[Pe] V.V.Petrov, *Sums of Independent Random Variables*, Springer-Verlag, 1975.

[R] M.S.Raghunathan, *Discrete subgroups of Lie groups*, Springer Verlag, 1972.

[RN] F.Riesz and B.von Sz.Nagy, *Lecons d'Analyse Fonctionelle*, 5ème édition, Gauthier-Villars, Paris, 1968.

[SC] L.Saloff-Coste, *Analyse sur les groupes de Lie à croissance polynomiale*, Arkiv för Mathematik **28** (1990), no. 2, 315-331.

[Saf] M.V.Safonov, *Harnack's inequality for elliptic equations and the Hölder property of their solutions*, J. Soviet Math. **21** (1983), no. 2, 851-863.

[Saz] V.V.Sazonov, *Normal Approximation-Some Recent Advances*, Lecture Notes in Mathematics, vol. 879.

[S1] E.Stein, *Singular Integrals and Differentiability Properties of Functions*, Princeton Univ. Press, 1970.

[S2] E.Stein, *Topics in Harmonic Analysis Related to the Littlewood-Paley Theory*, Annals of Math. Study vol.63, Princeton Univ. Press, 1970.

[S3] E.Stein, *Harmonic Analysis:Real-Variable Methods, Orthogonality and Oscillatory Integrals*, Princeton Univ. Press, 1993.

[Str] D.W.Stroock, *Diffusion semigroups corresponding to uniformly elliptic divergence form operators*, Seminaire de probabilitésXXII (P.A.Meyer and M.Yor, eds.), Lecture Notes in Mathematics, vol. 1321, 1988, pp. 316-347.

[Va] V.S.Varadarajan, *Lie groups, Lie algebras and their representations*, Springer-Verlag, 1984.

[V1] N.Th.Varopoulos, *Analysis on nilpotent groups*, J.Funct. Analysis **66** (1986), 406-431.

[V2] N.Th.Varopoulos, *Fonctions harmoniques positives sur les groupes de Lie*, C. R. Acad. Sci. Paris serie I **304** (1987), no. 17, 519-521.

[V3] N.Th.Varopoulos, *Analysis on Lie groups*, J. Funct. Anal. **76** (1988), 346-410.

[V4] N.Th.Varopoulos, *Small time Gaussian estimates of heat diffusion kernels I: The semigroup technique*, Bull.Sc.Math., 2e série **113** (1989), 253-257.

[V5] N.Th.Varopoulos, *Small time Gaussian estimates of heat diffusion kernels II: The theory of large deviations*, J. Funct. Anal. **93** (1990), 1-33.

[V6] N.Th.Varopoulos, *Wiener-Hopf theory and nonunimodular groups*, J. Funct. Anal. **120** (1994), 467-483.

[V7] N.Th.Varopoulos, *The heat kernel on Lie groups*, Rev. Mat. Iberoamericana **12** (1996), 147-186.

[V8] N.Th.Varopoulos, *Analysis on Lie groups*, Rev. Mat. Iberoamericana **12** (1996), 791-917.

[V9] N.Th.Varopoulos, *Diffusion on Lie groups*, Can. J. Math. **46** (1994), 1073-1092.

[VSC] N.Th.Varopoulos, L.Saloff-Coste and Th.Coulhon, *Analysis and Geometry on Groups*, Cambridge Tracts in Mathematics, 1993.

[Z] V.V.Zhikov, *A spectral approach to the asymptotic problems of diffusion*, Diff. Ouravnenia **25** (1989), no. 1, 44-55; English trasl., Differ. Equations **265** (1989), no. 1, 33-39.

Editorial Information

To be published in the *Memoirs*, a paper must be correct, new, nontrivial, and significant. Further, it must be well written and of interest to a substantial number of mathematicians. Piecemeal results, such as an inconclusive step toward an unproved major theorem or a minor variation on a known result, are in general not acceptable for publication. Papers appearing in *Memoirs* are generally longer than those appearing in *Transactions*, which shares the same editorial committee.

As of September 30, 2001, the backlog for this journal was approximately 7 volumes. This estimate is the result of dividing the number of manuscripts for this journal in the Providence office that have not yet gone to the printer on the above date by the average number of monographs per volume over the previous twelve months, reduced by the number of volumes published in four months (the time necessary for preparing a volume for the printer). (There are 6 volumes per year, each containing at least 4 numbers.)

A Consent to Publish and Copyright Agreement is required before a paper will be published in the *Memoirs*. After a paper is accepted for publication, the Providence office will send a Consent to Publish and Copyright Agreement to all authors of the paper. By submitting a paper to the *Memoirs*, authors certify that the results have not been submitted to nor are they under consideration for publication by another journal, conference proceedings, or similar publication.

Information for Authors

Memoirs are printed from camera copy fully prepared by the author. This means that the finished book will look exactly like the copy submitted.

The paper must contain a *descriptive title* and an *abstract* that summarizes the article in language suitable for workers in the general field (algebra, analysis, etc.). The *descriptive title* should be short, but informative; useless or vague phrases such as "some remarks about" or "concerning" should be avoided. The *abstract* should be at least one complete sentence, and at most 300 words. Included with the footnotes to the paper should be the 2000 *Mathematics Subject Classification* representing the primary and secondary subjects of the article. The classifications are accessible from `www.ams.org/msc/`. The list of classifications is also available in print starting with the 1999 annual index of *Mathematical Reviews*. The Mathematics Subject Classification footnote may be followed by a list of *key words and phrases* describing the subject matter of the article and taken from it. Journal abbreviations used in bibliographies are listed in the latest *Mathematical Reviews* annual index. The series abbreviations are also accessible from `www.ams.org/publications/`. To help in preparing and verifying references, the AMS offers MR Lookup, a Reference Tool for Linking, at `www.ams.org/mrlookup/`. When the manuscript is submitted, authors should supply the editor with electronic addresses if available. These will be printed after the postal address at the end of the article.

Electronically prepared manuscripts. The AMS encourages electronically prepared manuscripts, with a strong preference for $\mathcal{A}_{\mathcal{M}}\mathcal{S}$-LaTeX. To this end, the Society has prepared $\mathcal{A}_{\mathcal{M}}\mathcal{S}$-LaTeX author packages for each AMS publication. Author packages include instructions for preparing electronic manuscripts, the *AMS Author Handbook*, samples, and a style file that generates the particular design specifications of that publication series. Though $\mathcal{A}_{\mathcal{M}}\mathcal{S}$-LaTeX is the highly preferred format of TeX, author packages are also available in $\mathcal{A}_{\mathcal{M}}\mathcal{S}$-TeX.

Authors may retrieve an author package from e-MATH starting from `www.ams.org/tex/` or via FTP to `ftp.ams.org` (login as `anonymous`, enter username as password, and type `cd pub/author-info`). The *AMS Author Handbook* and the *Instruction Manual* are available in PDF format following the author packages link from `www.ams.org/tex/`. The author package can be obtained free of charge by sending email to `pub@ams.org` (Internet) or from the Publication Division, American Mathematical Society, P.O. Box 6248, Providence, RI 02940-6248. When requesting an author package, please specify $\mathcal{AMS}$-LaTeX or $\mathcal{AMS}$-TeX, Macintosh or IBM (3.5) format, and the publication in which your paper will appear. Please be sure to include your complete mailing address.

Sending electronic files. After acceptance, the source file(s) should be sent to the Providence office (this includes any TeX source file, any graphics files, and the DVI or PostScript file).

Before sending the source file, be sure you have proofread your paper carefully. The files you send must be the EXACT files used to generate the proof copy that was accepted for publication. For all publications, authors are required to send a printed copy of their paper, which exactly matches the copy approved for publication, along with any graphics that will appear in the paper.

TeX files may be submitted by email, FTP, or on diskette. The DVI file(s) and PostScript files should be submitted only by FTP or on diskette unless they are encoded properly to submit through email. (DVI files are binary and PostScript files tend to be very large.)

Electronically prepared manuscripts can be sent via email to `pub-submit@ams.org` (Internet). The subject line of the message should include the publication code to identify it as a Memoir. TeX source files, DVI files, and PostScript files can be transferred over the Internet by FTP to the Internet node `e-math.ams.org` (130.44.1.100).

Electronic graphics. Comprehensive instructions on preparing graphics are available at `www.ams.org/jourhtml/graphics.html`. A few of the major requirements are given here.

Submit files for graphics as EPS (Encapsulated PostScript) files. This includes graphics originated via a graphics application as well as scanned photographs or other computer-generated images. If this is not possible, TIFF files are acceptable as long as they can be opened in Adobe Photoshop or Illustrator. No matter what method was used to produce the graphic, it is necessary to provide a paper copy to the AMS.

Authors using graphics packages for the creation of electronic art should also avoid the use of any lines thinner than 0.5 points in width. Many graphics packages allow the user to specify a "hairline" for a very thin line. Hairlines often look acceptable when proofed on a typical laser printer. However, when produced on a high-resolution laser imagesetter, hairlines become nearly invisible and will be lost entirely in the final printing process.

Screens should be set to values between 15% and 85%. Screens which fall outside of this range are too light or too dark to print correctly. Variations of screens within a graphic should be no less than 10%.

Inquiries. Any inquiries concerning a paper that has been accepted for publication should be sent directly to the Electronic Prepress Department, American Mathematical Society, P. O. Box 6248, Providence, RI 02940-6248.

Editors

This journal is designed particularly for long research papers, normally at least 80 pages in length, and groups of cognate papers in pure and applied mathematics. Papers intended for publication in the *Memoirs* should be addressed to one of the following editors. In principle the Memoirs welcomes electronic submissions, and some of the editors, those whose names appear below with an asterisk (*), have indicated that they prefer them. However, editors reserve the right to request hard copies after papers have been submitted electronically. Authors are advised to make preliminary email inquiries to editors about whether they are likely to be able to handle submissions in a particular electronic form.

Selected Titles in This Series

(Continued from the front of this publication)

710 **Brian Marcus and Selim Tuncel,** Resolving Markov chains onto Bernoulli shifts via positive polynomials, 2001

709 **B. V. Rajarama Bhat,** Cocylces of CCR flows, 2001

708 **William M. Kantor and Ákos Seress,** Black box classical groups, 2001

707 **Henning Krause,** The spectrum of a module category, 2001

706 **Jonathan Brundan, Richard Dipper, and Alexander Kleshchev,** Quantum Linear groups and representations of $GL_n(\mathbb{F}_q)$, 2001

705 **I. Moerdijk and J. J. C. Vermeulen,** Proper maps of toposes, 2000

704 **Jeff Hooper, Victor Snaith, and Min van Tran,** The second Chinburg conjecture for quaternion fields, 2000

703 **Erik Guentner, Nigel Higson, and Jody Trout,** Equivariant E-theory for C^*-algebras, 2000

702 **Ilijas Farah,** Analytic guotients: Theory of liftings for quotients over analytic ideals on the integers, 2000

701 **Paul Selick and Jie Wu,** On natural coalgebra decompositions of tensor algebras and loop suspensions, 2000

700 **Vicente Cortés,** A new construction of homogeneous quaternionic manifolds and related geometric structures, 2000

699 **Alexander Fel′shtyn,** Dynamical zeta functions, Nielsen theory and Reidemeister torsion, 2000

698 **Andrew R. Kustin,** Complexes associated to two vectors and a rectangular matrix, 2000

697 **Deguang Han and David R. Larson,** Frames, bases and group representations, 2000

696 **Donald J. Estep, Mats G. Larson, and Roy D. Williams,** Estimating the error of numerical solutions of systems of reaction-diffusion equations, 2000

695 **Vitaly Bergelson and Randall McCutcheon,** An ergodic IP polynomial Szemerédi theorem, 2000

694 **Alberto Bressan, Graziano Crasta, and Benedetto Piccoli,** Well-posedness of the Cauchy problem for $n \times n$ systems of conservation laws, 2000

693 **Doug Pickrell,** Invariant measures for unitary groups associated to Kac-Moody Lie algebras, 2000

692 **Mara D. Neusel,** Inverse invariant theory and Steenrod operations, 2000

691 **Bruce Hughes and Stratos Prassidis,** Control and relaxation over the circle, 2000

690 **Robert Rumely, Chi Fong Lau, and Robert Varley,** Existence of the sectional capacity, 2000

689 **M. A. Dickmann and F. Miraglia,** Special groups: Boolean-theoretic methods in the theory of quadratic forms, 2000

688 **Piotr Hajłasz and Pekka Koskela,** Sobolev met Poincaré, 2000

687 **Guy David and Stephen Semmes,** Uniform rectifiability and quasiminimizing sets of arbitrary codimension, 2000

686 **L. Gaunce Lewis, Jr.,** Splitting theorems for certain equivariant spectra, 2000

685 **Jean-Luc Joly, Guy Metivier, and Jeffrey Rauch,** Caustics for dissipative semilinear oscillations, 2000

684 **Harvey I. Blau, Bangteng Xu, Z. Arad, E. Fisman, V. Miloslavsky, and M. Muzychuk,** Homogeneous integral table algebras of degree three: A trilogy, 2000

683 **Serge Bouc,** Non-additive exact functors and tensor induction for Mackey functors, 2000

For a complete list of titles in this series, visit the AMS Bookstore at **www.ams.org/bookstore/**.